VGM Opportunities Series

OPPORTUNITIES IN
BIOLOGICAL SCIENCE
CAREERS

Charles A. Winter

Revised by
Kathleen M. Belikoff

Foreword by
Charles M. Chambers, Ph.D.
Former Executive Director
American Institute of Biological Sciences

VGM Career Horizons
NTC/Contemporary Publishing Group

Library of Congress Cataloging-in-Publication Data

Winter, Charles A., 1902–
    Opportunities in biological science careers / Charles A. Winter;
foreword by Charles M. Chambers. — Rev. / by Kathleen Belikoff.
        p.   cm. — (VGM opportunities series)
    ISBN 0-8442-2300-X (cloth). — ISBN 0-8442-2301-8 (pbk.)
    1. Biology—Vocational guidance.   I. Belikoff, Kathleen.
II. Title.   III. Series.
QH314.W525   1998
570'.23—dc21   97-44573
                                                            CIP

Cover Photo Credits:

Top left, U.S. Fish and Wildlife Service; top right, Pedro Ramirez, Jr./U.S. Fish and Wildlife Service; bottom left and right, John and Karen Hollingsworth/U.S. Fish and Wildlife Service.

Published by VGM Career Horizons
An imprint of NTC/Contemporary Publishing Company
4255 West Touhy Avenue, Lincolnwood (Chicago), Illinois 60646–1975 U.S.A.
Copyright © 1998 by NTC/Contemporary Publishing Company
All rights reserved. No part of this book may be reproduced, stored in a retrieval
system, or transmitted in any form or by any means, electronic, mechanical,
photocopying, recording, or otherwise, without the prior permission of
NTC/Contemporary Publishing Company.
Printed in the United States of America
International Standard Book Number: 0-8442-2300-X (cloth)
                                                    0-8442-2301-8 (paper)

15   14   13   12   11   10   9   8   7   6   5   4   3   2

DEDICATION

To Greta

CONTENTS

ABOUT THE AUTHORS

Dr. Charles A. Winter is a former research worker in pharmacology at the Merck Institute for Therapeutic Research. At the time of his retirement, he accepted appointment as professor of pharmacology at Woman's Medical College in Philadelphia and also retained a consultantship with his former employer.

After receiving the A.B. degree in biology at Southwestern College, Winfield, Kansas, he studied zoology at the University of California and at Johns Hopkins. He then accepted appointment as instructor of biology at the University of Buffalo. While teaching at Buffalo, he took graduate work in physiology, receiving his Ph.D. He subsequently taught physiology at the University of Iowa and University of Oklahoma, before devoting full time to research at Merck Institute.

Dr. Winter is the author of more than a hundred scientific publications and has lectured at many international meetings in the United States and abroad. His memberships in scientific societies have included the American Association for the Advancement of Science, the New York Academy of Sciences, the Royal Society of Medicine in London, the American Physiological Society, the American Society for Pharmacology and Experimental Therapeutics, and others. He is a former chair of the Gordon Research Conference on Medicinal Chemistry.

Dr. Winter was called out of retirement to serve as acting director of research in Laboratoires Chibret in France. Upon his return from France, he settled in the state of Washington to be with his family, where he received appointment as consultant to the Department of Pharmacology at the University of Washington and visiting professor in the

Department of Biological Sciences at the University of California in Santa Barbara.

Kathy Belikoff personifies the idea that career opportunities abound for people who are willing to work their way up. Her own career in health care began when she took a temporary job as a medical records file clerk. She became a hospital administrator at Presbyterian Medical Center in Philadelphia and is now a health care consultant.

Ms. Belikoff grew up in suburban Baltimore and began her writing career as editor-in-chief of the Hereford High School student newspaper. She honed her writing skill as an English major at Hood College and pursued a graduate degree in education and information science at Towson State University. Her most recent publications are about her observations on the changing health care scene, including a chapter in *Using Hospital Space Profitably* about converting unused hospital space into restaurants, research labs, and other revenue producing facilities.

Until recently she was information specialist for the Scheie Eye Institute and in that role worked with all levels of clinical staff, researchers, administrators, medical and nursing students, and patients.

Working in a teaching hospital has been one of the most gratifying experiences in her career. Her special interest in adult education and the use of computers in literacy and vocational training began in graduate school, when she developed a media center at Springfield Hospital for institutionalized mentally ill adults who were preparing for reentry into the community.

FOREWORD

Biology is more than simply collecting specimens and observing them through a microscope. It is concerned with the development, structure, function, environment, and interactions of plants, animals, and microorganisms. An inquisitive nature about living things is ample motivation to consider a career in biology.

A wide range of interests is embodied within the biological sciences. Relationships between organisms and their environments are examined by ecologists. Researchers in the field of systematics will be busy for many years classifying the millions of species yet to be examined. Geneticists focus their studies on DNA replication and enzyme synthesis. Overlapping among the biosciences leads to even more specialized fields, such as plant physiology and forest entomology.

Workplaces are even more varied. Universities, zoos, hospitals, research foundations, and the government are just a few of the employers that hire biologists. In addition, "laboratories" may range from tropical forests to the Antarctic, from outer space to the ocean depths.

Before deciding to concentrate in a specialized field, however, you should try to gain an overall understanding of biology. Many environmental pollution problems can be solved only by applying biological principles. Health care specialists must understand the biology of the body before they can effectively treat a disease. Many college students declare a major in biology and later choose to specialize in areas such as botany, zoology, or ecology.

Career options in biology will continue to increase well into the twenty-first century, as new fields emerge to respond to our growing global problems. For example, a stable food supply for the world's increasing population may be provided through future discoveries in aquaculture or biotechnology, and many diseases, such as AIDS, may be cured and prevented using genetic engineering. *Opportunities in Biological Science Careers* provides a thorough presentation of the varied choices anyone with an interest in biology will want to explore.

I urge every individual seeking an exciting, challenging, and worthwhile career to consider the many opportunities available in the biological sciences.

Charles M. Chambers, Ph.D.
Former Executive Director
American Institute of Biological Sciences
Washington, DC

PREFACE

Changes are occurring so rapidly in our culture that it is a challenge to find an occupation that will not become obsolete before the end of one's productive years. The career you choose should be one that you enjoy. Whether you choose the biological sciences or a related field, you will find that many career choices await you. Biologists are happy people! The rewards of their work are not only a comfortable living, but also the knowledge that they leave the earth a better place.

This fifth edition of *Opportunities in Biological Science Careers* incorporates the many changes that have occurred in the field over the last few years. The science of biology has grown in many ways including the introduction of computers, the popularization of science in the media, the advances in genetic engineering, and the worldwide interest in ecology. A recent survey conducted by the National Science Foundation found that two-thirds of the people interviewed thought that science was very worthwhile. However, the survey also showed that only one in four had some knowledge of science and a troubling 98 percent did not understand the scientific process or how research is funded.

We hope that by reading this new edition of *Opportunities in Biological Science Careers,* you will choose to pursue a career in the biological sciences. If you do not, we hope that you will come away more informed and in support of scientific method as a necessity to the survival and improvement of all living things.

C. A. W.
Redmond, Washington

CHAPTER 1

WHY BECOME A BIOLOGIST?

If you have a curiosity about the nature of the world and the life upon it, you have one of the main attributes for success in the challenging career of biological sciences. If you are somewhat of an idealist, inspired by the prospect of improving the lot of your fellow human beings, there are many areas of biology that you will find gratifying. If you are of a practical turn of mind and are concerned about choosing a reasonably secure job that will afford you and your family a comfortable living, then a career in biology offers a number of choices.

One frequently hears biologists—or other scientists—marvel at their good luck in getting paid for something that is so much fun. Most would do it even if they no longer had to work for a living. Science is a way of life and scientists, in general, are happy people. The motivation for following science as a career comes from within, and, for many, the urge to pursue science is as strong as a religious calling. And like a religious calling, the scientist's greatest reward is knowing that he or she has had a positive impact on mankind and the other creatures of the earth.

UNDERSTANDING THE "SCIENCE" OF BIOLOGY

The immediate aim of science, generally speaking, is *not* usefulness. Like a beautiful painting or a melodious musical composition, the fruits of science can enrich our lives. According to *Webster's Dictionary* "science" is "the observation, identification, description, experimental investigation and theoretical explanation of natural phenomenon." For many people, science is a quest that challenges us to pursue the spirit of

1

inquiry, the enthusiasm for exploration, the habit of rigorous analysis, the inquisitive outlook, the search for truth, and the willingness to discard pet notions once they are found to be mistaken. The scientist personifies our highest ideals of total objectivity and scrupulous integrity. As the biologist Konrad Lorenz put it, "It is a good morning's exercise to discard a pet hypothesis every day before breakfast: it helps to keep us young." Such a change of mind would be very disturbing to the average adult.

There is confusion in the minds of many people as to the distinction between science and technology or invention. The confusion is understandable. We are accustomed to frequent changes in our way of life brought about by new discoveries and inventions. For example, new drugs help to control diseases; new varieties of plants and animals increase agricultural yields; new processing methods bring us new foods (though not always improving on taste or nutritional value); and transportation speeds up until we can move through the air faster than sound can travel. Such innovations often are regarded as advances of science, and many people believe that invention is the object of science. That is not really true, even though the inventions would not be possible without the scientific discoveries upon which they are based. Many very fine inventors are not scientists at all; while most scientists have never invented anything of practical value.

There is, then, a distinction between science and technology. Science seeks to understand the forces of nature, and the scientist's reward is the joy of discovery and a contribution to knowledge. Technology, on the other hand, puts this knowledge to practical use. In some instances, the distinction becomes somewhat subtle. We sometimes hear about *pure* or *basic* research as distinguished from *applied* research, and there are arguments about whether scientifically trained people are truly scientists if they use their talents for making discoveries that they hope will lead to a practical result. Basic investigations often lead to practical results—sometimes years after the basic discovery has been made.

When the British biologist, Alexander Fleming, discovered that a substance produced by the bread mold *Penicillium* could kill bacteria, he was engaged in basic research, but in the end his inquiry led to the important practical result of the isolation of penicillin. This work ushered in a whole new era in medicine—the antibiotic era. Penicillin would not have become a useful product, however, without a large amount of practical research and the application of skilled technology.

Hence, science and technology often work closely together, and enlightened industrial leaders have dedicated more and more of their funds to basic science. Today, much important basic work in the life sciences as well as in other sciences is performed in the laboratories of industry.

Another distinguishing feature of scientists is that they don't have to pretend to know something in the field that they don't know. Scientists may guess what the answer will be and may have ideas about how to go about finding the answer; but if they don't know, it is not shameful to say so. A large scientific gathering was addressed by the then-president of the American Association of the Advancement of Science. Regarding a difference of opinion between himself and another eminent worker in the same field, he said, "Of one thing I am quite sure; neither of us knows anything about it." Louis B. Flexner once said of his work, "Although I have confidence in the observations, I am far from wedded to the interpretations, which, in my opinion, badly need further work to test them."

How refreshing it would be if many people in other fields were to adopt a similar attitude! If you choose to become a scientist, you will be privileged to meet many people with such attributes, and you will be proud that you have chosen a career in science.

Biology is a group of sciences, rather than a single science. But no matter how the daily activities of one biologist may differ from those of another, all bioscientists are concerned one way or another with living creatures. All living things have certain properties in common, whether they be plants, animals, or microorganisms. All require sustenance derived from materials absorbed from the environment. They all grow and develop, and all respond in predictable ways to the environment. All of them depend upon reproduction to perpetuate their kind. All of these processes are governed by natural laws that are the same for all. Biology deals with all of these things—and more.

Many of the problems of modern civilization can be solved only by the application of biological knowledge. Examples of these problems are easy to find: pollution of air, water, or soil; the disposal of wastes; the effects of overpopulation and crowding; the use of drugs; the aging of the population; and the spread of AIDS (acquired immune deficiency syndrome). Improvement of the food supply and elimination of heritable diseases also are challenges to the biologist. Biology is also involved in solving problems associated with descent into the depths of the sea and ascent to the highest mountains—and even into outer space.

You can readily appreciate, then, that biology is much more than taking specimens of plants and animals into the laboratory and looking at them and describing what you see. It may interest you at this point to take a brief look at the types of biology and the range of careers available to those trained in bioscience.

A NICHE FOR EVERY INTEREST

The classification of biologists as botanists, zoologists, or microbiologists gives little hint of the wide variety of biological disciplines that exist. For example, a microbiologist whose primary interests involve bacteriology may perform very different duties from those carried out by a microbiologist working in virology.

One kind of zoologist, the taxonomist or systematist, specializes in the identification, description, and classification of animals and determines how a given species fits into the scheme of nature. Seldom will the interests of a zoological systematist range over the whole animal kingdom; he or she is more likely to specialize to some extent. For example, a person who mainly studies mammals is a mammologist; reptiles, a herpetologist; insects, an entomologist; fish, an ichthyologist; protozoa, a protozoologist; and birds, an ornithologist. Botanists and microbiologists can be systematists, too. A diminishing breed is the generalist, a systematist whose interest range across the boundaries between plants and animals.

Bioscientists of several categories are fascinated by the structural details of organisms—they are anatomists or morphologists. If the microscopic structure of tissues is the object of the morphologist's scrutiny, he or she is a histologist. Still more detailed study of the structure within cells makes the morphologist a cytologist. If this sounds like morphology is a lifeless and dull subject, the impression is far from the truth. The cytologist, for example, is keenly aware of the relationship between the structural components of a cell and cellular functions. So, cell morphology becomes a background for other branches of cellular biology, such as cellular physiology or biochemistry. The molecular biologist is preoccupied with an even more intimate examination of the ultimate mechanisms within a cell.

From this brief survey, you can see that the interests of bioscientists range all the way from a systematic examination of whole groups of plants or animals (or both) down to the level of the molecules that make up cells. But that is only the beginning! No group of plants or animals exists by itself. All groups and individuals interact with one another and with their environment, and each has a niche to fill in the scheme of nature. Its whole surroundings form an ecosystem, so another biological science, ecology, is born. Patterns of behavior stimulated in each species of animal by its encounters with the world around it provide the subject matter for the science of ethology.

The laws governing heredity occupy the attention of the science of genetics. Geneticists may apply mathematics, anatomy, biochemistry, or a host of other disciplines in their efforts. The lines between different biological vocations are often rather poorly defined, and there is much overlap. For example, if a life scientist is interested in the genetics of bacteria, is the scientist a geneticist or a bacteriologist? If the favorite subject is the embryonic development within the egg of a reptile, is the scientist an embryologist or a herpetologist? Is the botanist who writes papers on diseases of trees caused by fungi a mycologist (student of fungi), a dentdrologist (expert on trees), or a plant pathologist (concerned with diseases of plants)? In each case, he or she is a bit of each and can be classified according to what the primary interests are—and this can change during different stages of the scientist's career.

Wildlife biologists are concerned with wild animals, fish, and fowl, and marine biologists are fascinated by the richness of life in the sea, but either may concentrate on taxonomy, physiology, ecology, or any of several other specialties.

All biological disciplines are basic sciences, but each one also has practical applications. Thus, agriculture and forestry make wide use of plant and animal sciences. Those biosciences having important applications in health and diseases of human beings or other animals are commonly referred to as biomedical sciences. They include certain aspects of biosciences we have already mentioned, such as anatomy, physiology, biochemistry, and microbiology, as well as such disciplines as biophysics, nutrition, pharmacology, and pathology. All of these will be discussed in succeeding chapters. When the public press mentions "the advances of medical science," the reference is usually to reports of research in one or more of these life sciences.

Some of the newer biological specialties that have emerged in recent years and may expand significantly in the future include genetic engineering or biotechnology, involved in manipulations of the units of heredity; biomedical engineering, concerned with replacement of faulty organs of the body; gnotobiotics, dealing with breeding and growing animals in germ-free environments; teratology, the study of abnormalities, which has become important with our increasing exposure to harmful environmental influences—teratologists are widely employed in laboratories of the chemical and drug industries; aquaculture, an increasingly widespread technique for growing foodstuffs in water, especially high protein foods such as fish; and hydroponics, growing plants without soil in an aqueous medium. Some of these specialties have been around for many years but are assuming greater and greater importance with the increase in population and decrease in available arable land. Other biological specialties that are expected to be in increasing demand include paramedical personnel who assist in health care—medical technicians, physician assistants, and others.

During a lifetime career, a bioscientist may engage in teaching, research, administration, service work, or industrial production. If you choose to become a biologist, you may specialize in one of these activities or a combination of two or more, or you may switch from one to another at various stages of your career. Your employer may be an educational institution, governmental installation, private research foundation, zoo, botanical garden, aquarium, arboretum, natural history museum, or an astonishingly wide variety of industrial or business concerns. Few biologists find life dull.

THE CONTINUING EVOLUTION OF THE BIOLOGICAL SCIENCES

The Ancients

Like nearly everything else in our Western culture, biology can trace its origins back to the ancient Greeks and Romans. In the fourth century B.C., Aristotle classified plants and animals, and his pupil, Theophrastus, wrote the first botany book. Some of their notions seem strange to us. Aristotle's theory that the same kinds of animals would be found on the same parallel of latitude around the world so misled Columbus that he was disappointed at not finding elephants in Haiti. Hippocrates is still revered for laying the foundations of medical science. In the first century A.D., Seneca made observations aided by magnifying glasses consisting of globules filled with water.

Biology Awakens after the Renaissance

The biological lore accumulated by the ancients was forgotten in Europe for more than a thousand years. The ancient records kept alive by Arab and Jewish scholars in the Middle East were brought back by returning Crusaders. Even after the Renaissance, progress was very slow by modern standards. During the next three hundred years, only a few names stand out—Andreas Vesalius, the "father of anatomy;" William Harvey, who discovered the circulation of the blood; and Robert Hooke, who introduced the concept of cells as units of biological structure. Seneca's crude ancestor to the microscope was not improved upon until the late seventeenth century when a Dutchman, Anton van Leeuwenhoek, devised the first microscope and discovered the existence of animals and bacteria too small to be seen with the unaided eye.

The eighteenth century witnessed several advances in what we would now call biochemistry. Joseph Priestley, an Englishman, discovered that green plants give off oxygen, while two Frenchmen, Antoine Lavoisier and Pierre Simon de Laplace, pointed out that respiration of animals is a form of combustion, like the burning of wood. The great Swedish naturalist Carl von Linne, commonly known by the Latin form of his name, Carolus Linnaeus, devised the scientific method of classifying and naming plants and animals that is in use to this day.

The Nineteenth Century—The Modern Age Arrives

The immense changes in our way of life that came during the nineteenth century also saw revolutionary new ideas in biology. All living organisms are made up of cells, theorized two Germans, Matthias Schleiden and Theodor Schwann. Another German, Justus von Liebig, discovered that plants synthesize organic compounds from carbon dioxide in the atmosphere and nitrogen from the soil. Louis Pasteur of France pioneered in microbiology and proved that living organisms do not arise spontaneously from nonliving matter—all life comes from life. His discoveries and those of the German bacteriologist Robert Koch paved the way for many advances in the treatment of disease, including aseptic surgery. The principles of heredity were also discovered during the nineteenth century by an Austrian monk, Gregor Mendel, though his work lay forgotten until the beginning of the twentieth century.

No event in biological science in the nineteenth century had greater impact than the publication in 1859 of Charles Darwin's *The Origin of Species,* setting forth the theory that species of animals and plants gradually change over long periods of time because of natural selection. The idea that species could be transformed was not entirely new. Anaximander in the sixth century B.C. had proposed that one species could evolve into another. More than a century before Darwin's publication, the French naturalist Georges Louis Leclerc de Buffon asserted that species could change. Charles Darwin's grandfather, Erasmus Darwin, thought that species could be transformed by environmental influences. The Frenchman Jean Baptiste de Lamarck also had emphasized the fundamental unity of all life and the capacity of species to vary.

Even in the idea that natural selection brought about evolutionary changes, Darwin was not alone. Alfred Russel Wallace came to identical conclusions at the same time; indeed, the initial public disclosure of the theory was in a joint presentation of papers by Darwin and Wallace to the Linnaean Society in London. The subsequent publications in greater detail by Darwin, including *The Origin of Species* and *The Descent of Man,* have led to the association of his name, rather than Wallace's, with the evolution theory.

The Twentieth Century—and Beyond

In looking back on the developments of the twentieth century, we might call it the era of the "popularization" of science. Particularly after World War II, science came out of the laboratory and into people's homes improving almost every aspect of their lives. The introduction of television brought glamour to science. Thanks to television, who could ever forget seeing the first space walk on the moon, and many baby boomers grew up dreaming of joining Jacques Cousteau and the adventurous band of oceanographers on the *Calypso.* In addition to television, the computer also revolutionized and popularized biology and the sciences. Microchip technology has been incorporated into almost every aspect of biological research, and access to the Internet superhighway via the personal computer has provided researchers and the general public worldwide access to scientific information that was only available in university libraries. In the twentieth century, as the demand for science-driven advances increased, so did the educational and employment opportunities for scientists, including women. Ecology, one of the fastest growing areas of biology, attracted some of the most famous women in science. Many people credit Rachel Carson, a marine biologist by training, with the launching of the ecology movement. Carson's best-selling book, *Silent Spring,* is considered one of the most influential books of the last fifty years. Her indictment on the use of pesticides taught people to think about being connected to the earth in an interdependent way. Another conservationist, Jane Goodall, captured the hearts and minds of the public with her research on the behavior of chimpanzees of the Gombe Game Reserve in Africa. Through her films, books, and *National Geographic* articles, Goodall helped millions of people understand the relationship between all creatures and inspired many to commit to preserving the delicate balance of the world's ecology.

Genetics is one of the fastest growing areas in science and will continue to occupy the attention of the public and scientific community well into the twenty-first century. Early in the twentieth century, evolutionary changes within a species were essentially changes in the frequency of the occurrence of specific "genes" within a population. Genes, the units of heredity, interact with each other and with the environment to produce the manifestations of inherited characteristics. The gene theory was formulated early in this century by the American biologist Thomas Hunt Morgan, who discovered that genes are located on cell structures called chromosomes. In the 1940s, scientists found that genes consist of segments of DNA (deoxyribonucleic acid). In 1953, two biochemists, James D. Watson of the United States and Francis H. C. Crick of England, described the structure of DNA as a twisted-ladderlike *double helix.* Since then, so much has been learned about how to handle DNA that some fear this new knowledge gives biologists altogether too much potential power to produce whole new species of dangerous disease-producing bacteria and perhaps other monstrous forms, or maybe even control heredity in human beings. Biologists working in this field, aware of the potentialities, have met and agreed upon certain rules aimed at guiding research along lines that will not be harmful to humankind. In addition, the potential benefits of understanding the complexities of DNA and of work in genetic engineering have already been demonstrated. For example, researchers have succeeded in using genetically engineered bacteria to produce human insulin for treatment of diabetes.

Examples of biological discoveries will be given throughout this book. Exciting new findings are lying in wait for those with prepared minds. Many will be made by people now barely beginning the study of bioscience and possibly by some who have not yet made up their minds whether they really want to be biologists. Perhaps this book will help you make that decision.

OPPORTUNITIES FOR EVERYONE

Recently biology, unlike some of the other sciences, has been successful in attracting a diversified pool of aspiring scientists. According to the U.S. Department of Education, women now comprise more than half of the biology graduates in the United States, and the number of African American, Hispanic, and Native American students who major in biology has grown rapidly in the last decade.

This success can be attributed to the commitment of educational institutions, industry, and philanthropic organizations that have partnered to revolutionize biology education. In its recent landmark study, "Beyond Biology 101: The Transformation of Undergraduate Biology Education," the Howard Hughes Medical Institute reported on the "kaleidoscope of approaches" used by a number of colleges and universities to attract minority and nontraditional, especially older, students to the study of biology. The study cites the following rationale for the success of these programs:

> Just as biology has found unifying themes in the function of DNA, the process of natural selection, and the dynamics of ecosystems, so biology education has been uncovering the principles that underlie effective teaching and learning. Instructors are finding that students learn more when asked to engage in critical thinking and problem solving. Changes in the workplace are highlighting the need for people to be able to communicate their ideas and work in teams. Innovative programs are demonstrating how to involve a broader cross section of young people in the sciences, especially women and minorities…the rapid growth of biological knowledge is itself a powerful force of change. Instructors know that they can no longer cover everything; instead, they are increasingly focusing on concepts that cut across scientific fields. Biology is building bridges to chemistry, physics, mathematics, information science, and other disciplines, requiring that its students become well grounded

through the sciences. Classroom and laboratory instruction are emphasizing the flexibility that students need to master rapidly advancing specialties and techniques. And as biology itself assumes an increasing prominence among the sciences, biologists are realizing that their responsibilities have changed. They must produce not just the doctors and research scientists of the future but also a biologically literate citizenry.

The sciences also have been in the forefront of offering opportunities to the physically challenged. There is little concrete data on the exact number of disabled scientists probably because they and their employers are more interested in doing good science. In Chapter 2 you will be reading about what it takes to be a successful biologist. You'll notice that nowhere in this chapter do we mention being a perfect human specimen!

In 1975, the American Association for the Advancement of Science (AAAS) initiated its Project on Science, Technology and Disability. The project publishes a directory of more than 1,000 scientists and engineers with disabilities who are willing to share their experience with others. The AAAS group, corporations, and other organizations have made great strides in promoting science as a career for disabled persons. They have made progress in such areas as monitoring professional meetings to ensure that they are barrier-free, developing assistive devices technology for classrooms and labs, and science education from preschool to postsecondary schools. The National Science Foundation provides Facilitation Awards for investigators, staff, and student research assistants who require special assistance or equipment to work on projects funded by the NSF. More information about these services is included in the appendixes at the end of this book.

THE EMPLOYMENT OUTLOOK FOR BIOLOGISTS

According to the *Occupational Outlook Handbook,* which is published by the U.S. Department of Labor, the employment of biologists and medical scientists is expected to increase faster than the average for all occupations through the year 2005. The fields of biology that are expected to grow the most will be in genetic and biotechnical research, environmental protection, and biomedical research that deals with health issues such as AIDS, cancer, or the Human Genome project.

Historically, the number of new scientists joining the workforce has increased approximately 20 percent each decade. This trend is expected

to continue into the first part of the next century. Of the half million people in the United States today who call themselves scientists, there are
currently about 150,000 bioscientists. Of these about one-third are general biologists and two-thirds specialize in agriculture, forestry/
conservation, or the medical sciences.

As in many professions, scientists in all disciplines are acquiring a
more global perspective when considering employment opportunities.
Cutbacks in both federal and private industry funding are expected to
continue in the United States and Canada while research expenditures
are increasing in foreign countries such as Japan and Germany, where
scientists are needed to provide the basic research that is the foundation
for high-tech innovation.

Competition for top jobs anywhere in the world will be keen into the
twenty-first century. Job seekers with a doctoral degree in biology will
continue to have the competitive edge for the best paying positions. A
combination Ph.D. and M.D. is becoming desirable for biomedical
researchers who are qualified to follow their investigations from the laboratory bench to the clinical setting.

The employment outlook is expected to be positive for not only new
entrants into the job market, but also for biologists currently working in
the field. Because they are often involved in long-term research or grant
projects, biological and medical scientists are less likely to be laid off
than are other workers.

THE TANGIBLE AND INTANGIBLE REWARDS

The rewards of a career in the biological sciences are both tangible
and intangible. The tangible rewards include a safe, efficient working
environment and a compensation package that includes a salary and
benefits that will provide a comfortable living for you and your family.
Perhaps even more important are the intangible rewards of this career
choice. Throughout their careers scientists experience continuous satisfaction and pride in knowing that they have had a positive impact on
peoples' everyday lives, as well as the future of our entire planet. Many
biologists will tell you that they cannot believe that they get paid for
doing something that is too much fun to be called work!

Part of the enjoyment of being a biologist is working in laboratories
or classrooms or both, which are usually comfortable, clean, well

lighted, and adequately equipped. Many biologists choose more exotic work sites such as an experimental field station or a research vessel. Wherever a biologist works, he or she will usually find whatever office space, library facilities, and other services that are needed. These are generally supplied at the expense of the employer. Most will work in the company of others, and there will be stimulating interactions among fellow scientists, technicians, assistants, and others. For those few jobs involving the handling of hazardous materials, special training is available. Recent regulations from the Department of Occupational Safety and Health Administration (OSHA) require that all workers, including scientists, be well informed about the labeling, safe handling, and emergency spill procedures of any substance that can be toxic when opened or if there is prolonged exposure in the workplace.

Since competition for top scientists and technologists is keen, other features of the work environment or "quality of life" at work are often the deciding factors in choosing among employers. Biologists, like other corporate or educational professionals, are attracted to facilities that offer amenities such as access to personal and mainframe computers, onsite or nearby childcare, exercise facilities, window offices and attractive surroundings, cafeteria or snack centers, a nonsmoking environment, access and accommodation for physically challenged workers, places and events that promote comradery among employees, etc. And this environment, where casual dress (but not jeans) is usually considered suitable attire for under the lab coat or in front of the classroom, should be appealing to those who cannot envision themselves going to work each day in a suit and tie.

The other tangible rewards of a career in the biological sciences—pay and benefits—are meant to sustain the biologist's quality of life outside of the working environment. These rewards, often referred to as a "compensation package" include:

- salary, including bonuses and merit or cost-of-living raises
- health, dental, prescription, and vision benefits
- pension or portable (401K) retirement plans
- profit-sharing, stock options, and tax-sheltered investments
- direct-deposit banking and credit union
- disability and life insurance
- continuing education onsite for college degree, and allowances for conferences

- time off: vacation, holidays and personal, sick, maternity and other leaves of absence
- working hours: flexible schedule; telecommuting, overtime pay, compensatory time off
- transportation allowances: parking, train/bus pass reimbursement
- childcare or elder care facilities or allowances

While comparing the salary figures offered by two or more employers is somewhat straightforward, analyzing the value of a "cafeteria line" of flexible benefits can be confusing. Employers are constantly redesigning their recruitment packages to gain the competitive edge in hiring the best candidates. That is why it is important to ask a potential employer for a written description of salary and benefits being offered, and then carefully evaluate the entire compensation arrangement, to see if it will match the personal and professional needs of you and your family.

The salaries for biologists vary greatly depending on the level of education required and the scope of the job, the size and location of the employer, and the conditions of the labor marketplace. Also, as in other professions, the law of supply and demand determines how an individual is paid. If your talents are in demand and there are only a small supply of other people besides you who can perform the job, then you are in the driver's seat for commanding the salary you want. If your talents are not in demand or there is a large supply of people including you who can perform the job, then the employer is in the driver's seat for salary negotiation.

Fortunately, the demand for a small supply of biologists is growing and, therefore, the salaries are growing, too. Starting salaries for college graduates with a bachelor's degree range between $20,000 and $30,000. The average salary of biologists with various levels of education employed by the federal government is about $47,000. Salaries in private industry for similar positions are about 10 percent higher or in the low to mid $50,000 range. Only 10 percent of all biologists make in excess of $60,000.

Teachers and professors of biology are paid according to their level of education as well as the location and status of their institutions. Public secondary school teachers in the United States average about $37,000, with higher salaries averaging around $46,000 in major metropolitan areas. Almost all college and university professors must have Ph.D.s and are paid based on their faculty rank. The following are average salaries for faculty: instructor, $29,680; assistant professor, $39,050; associate professor, $47,040; and full professor, $63,450.

Although practically everyone needs and appreciates a good paycheck, to the average biologist the deepest satisfaction about his or her career comes from things no amount of money can buy. For example, for the high school or college teacher of bioscience, the thrill of watching young people with keen minds react to the excitement of science cannot be described, only experienced. The delight and wonderment that the young show in living things and the satisfaction that the teacher has in observing comprehension and enthusiasm appear on the faces of students are the rewards of teaching. But this is not all. Some of the students will choose bioscience as a career and extend the teacher's influence to yet another generation—and even those who do not choose a career in biological science may continue an interest in living things and an attitude toward life that will reflect the influence of good teaching.

Research work, administration, or any other kind of service in bioscience has intangible rewards no less than those of teaching. Think of the number of people you know to whom their daily work is "just a job." Also think how much of life people miss if they do not find pleasure and gratification in their day-to-day occupation! If there are any such people in biology, they are in the wrong profession. One of the most obvious things about life science is the enthusiasm most biologists have for their life's work. Not only is the work itself interesting, but most biologists take pretty seriously the idea that their efforts are worthwhile and that they can leave the world a little better than they found it. Perhaps this is one reason biologists are so much inclined to talk shop after hours, and maybe it explains why so many of them continue studying, thinking, and working after retirement.

CHAPTER 2

IS A CAREER IN BIOLOGY FOR YOU?

The personal qualities that improve one's chances for success in biology are desirable in any scientist. This does not mean that all biologists have to fit a pattern or that all biologists are alike. Biology, probably the most varied of all the sciences, is broad enough to challenge anybody possessing curiosity about the nature of the world and of its life. As you read this chapter, compare yourself to the profile of a successful biologist. Do these characteristics describe you and your interests? Consider the difficult ethical decisions facing today's scientists. As a biologist, did you think that you would only deal with clearly defined facts? Think again! By the end of this chapter, you will know if you have the "stuff" that successful biologists are made of.

PROFILE OF A SUCCESSFUL BIOLOGIST

Enthusiasm

The personal attribute that heads most lists of desirable traits for scientists is genuine liking for the subject. Do you enjoy studying biology in high school or college? Do you like to read about science in magazines and newspapers? Have you ever made a collection of natural objects or made a special study of any group of living creatures, such as flowers, trees, butterflies, or seashells? Do you like to attend scientific exhibits, or have you ever taken part in a science show? Do you find the performance of laboratory experiments an absorbing interest? Do you like to work with animals or with plants? If you can answer most of these questions with a "yes," you may have one of the most important qualifications for success in a biological career.

Intelligence

Don't let the heading of this paragraph give you the wrong idea that only those of extraordinary intelligence can be good biologists. Emphatically, one does not have to be a genius to succeed in science; individuals such as Einstein and Pasteur are few and far between. People with an extremely wide range of talents can find a place in the life sciences. As in any field of endeavor, of course, those with keen mental alertness and better-than-average intelligence stand a better chance of getting to the top.

The ability to reason both abstractly and concretely—that is, to solve such problems as algebraic equations and to understand the significance of new observations—will be useful, especially in research. When new facts are presented, one should have the ability to fit them in with things already known. At times, the new facts upset previous prejudices and notions; biologists must be prepared to adjust thinking accordingly. This capacity for intellectual growth is probably more important than is the possession of a high degree of native intelligence.

Curiosity

Others who have thought about the personal attributes of biologists have all given a prominent place to curiosity. Some years ago, the American Institute of Biological Sciences put it as follows: "This inquisitive outlook, this penetrating quest for the truth, is the most distinguishing characteristic of all biologists."

Advances in science have been made by those who *wondered* about things. This attribute is especially important for those who will enter biological research, but it surely is not out of place even in the performance of routine lab investigations or fieldwork. In addition, for the teacher of biology, the ability to foster an inquiring state of mind in one's pupils is a precious asset. If you wish to seek new frontiers of knowledge, there is no subject that will give you more satisfaction than biology. The quest for new knowledge takes biologists all over the world, from the icebound continent of Antarctica to the steaming Amazon rain forest, and from the tops of mountains to the floors of the deepest seas. Biologists are also in the forefront of the space program, and several biological experiments have been performed aboard the space shuttle. Bioscientists in many laboratories are devoted to the conquest of disease, even delving into the nature of life itself.

Open-Mindedness

We have already pointed out that a good biologist is willing to discard an old notion if it does not fit newly discovered facts. This is such an important qualification that it warrants having a section of its own in the present list. It should be made clear that there is a difference between being open-minded and being vacillating; one should not be blown about by every change in the wind. Moreover, it is the interpretation that changes, not the facts themselves. An observation made with sufficient care should be valid indefinitely, but the things seen by an observer may change with alterations—often very subtle and ill-defined alterations—in the conditions under which an observation was made. A good biologist forms opinions, not on the basis of preconceived notions, but on observed facts.

Precision

It will be useful to many biologists to be able to make observations with great accuracy and to detect minute differences between object or event. Two kinds of plants or two kinds of insects may look so much alike that the untrained observer would think they were the same, though they may be different species—that is, so different that they cannot interbreed. It may be equally important to recognize similarities. Small differences also may be important in laboratory results. If you feel that you cannot make such fine distinctions, it may be because you have never been called upon to do so; the ability to make accurate observations is a quality that can be learned, and every serious student of biology receives training in it.

Patience

When the great German biologist Paul Ehrlich sought to find a chemical compound that would destroy the parasite *Treponema* without killing the patient, he finally found it after trying 606 different substances. Although "606" has been long since superseded by better and safer remedies, its discovery was an important milestone in the history of biomedical science, for it initiated the era of chemotherapy—that is, the treatment of diseases with chemical compounds. Up to that time, the only agent that could be classified as chemotherapeutic was the ancient remedy quinine, which was extracted from the bark of a tree—cinchona—and used to kill malaria parasites. In that prechemotherapeutic age, Oliver Wendell Holmes, a physician as well as a poet, told the Massachusetts Medical Society: "If the whole *materia medica* as now used could be sunk to the bottom of the sea, it would be all the better for mankind—and all the worse for the fishes."

If Ehrlich and his assistants had lacked patience, they might have said, "We will try 600 substances and if none of them works, we'll give up." This great advance then would have had to await the appearance of a more patient investigator. Actually, Ehrlich was extraordinarily lucky, for most modern drugs are selected after trials of many thousands of substances. One of the modern antibiotics, terramycin, for example, was found only after the study of nearly 100,000 molds!

It is true not only in the field of drug therapy but in nearly all biology that most trails lead nowhere. One doesn't know that a trail is a blind alley until it has been explored, so one must follow a path for a while before starting all over again. Often the same manipulations and observations must be repeated over and over again with a minor modification each time until the true path is found. It is not commonly appreciated by the general public that success is the exception rather than the rule and that there is a great deal of trial and error associated with every great discovery. But most investigators take keen pleasure in the pursuit of knowledge and the joy of discovery—whether in the laboratory, the field, the classroom, or elsewhere—which more than compensates for the long period of testing and measuring and repeating that goes on beforehand.

Other Personal Attributes

The list of desirable attributes could be continued until it included everything we admire in people. There are, however, certain qualities that are especially helpful in particular kinds of jobs. Most biologists will need to exhibit a willingness to work with others and a genuine liking for people. These attributes are obviously desirable for teachers, but they are usually an asset in other kinds of work as well; the days of the "lone wolf" are pretty much in the past. Most work today is done in cooperation with others. In some types of laboratory work, manual dexterity would be a valuable asset, particularly when one must manipulate very small objects. The degree to which this is required varies greatly among different types of biological jobs.

Writing and speaking ability will help not only teachers but others as well. The higher one advances, the more important it becomes to be able to communicate with fellow scientists. Biologists can obtain an enviable reputation only if their fellow scientists understand what they are doing, and this is communicated in the form of papers published in professional journals and in lectures at meetings and conventions. To a considerable degree, an aptitude for speaking and writing can be learned by study and practice, so you should not be discouraged if you feel that you lack this aptitude. For that matter, there are many very fine jobs in bioscience that do not require public speaking or writing, but generally these are not the top-grade jobs in teaching, research, or administration.

Certain specialized attributes will be helpful in specific types of jobs. For example, some of the duties of the wildlife biologist may require considerable physical stamina and strength, with some competence in outdoor craft. Arduous outdoor activity, sometimes in remote areas, may be required. On the other hand, there are laboratory and office jobs in wildlife biology, too.

Some of the attributes listed above are somewhat special— particularly enthusiasm for the subject, curiosity, open-mindedness, and patience. Perhaps you feel uncertain as to whether you possess the requisite attributes. If so, you should not feel discouraged, for if you'll review the list above, you may notice that by and large they are attributes toward which anyone can aspire, and many of them can be acquired.

The chances are that if you have gotten this far reading this book and if you have had some courses in high school science, you already have a fair idea as to whether biological science appeals to you. If it does, and if you are willing to work faithfully, you should be able to succeed in a biological career and find immense joy and satisfaction in it. If you do decide to follow such a course, you will find that the rewards are great.

THE BIOLOGIST'S CONSCIENCE: GRAPPLING WITH THE BIOETHICS OF SCIENTIFIC RESEARCH

For the biologist, progress is a double-edged sword. With each advance, there is both the potential to do good and the potential to do harm. In a society where the expectation for a high level of technology clashes with the limited availability of resources, the boundaries between good and harm often are blurred.

In addition to the personal attributes already described in this chapter, the biologist must have a strong sense of what is right and what is fair. This is what constitutes ethical behavior in the world of research. It is also the greatest intangible reward for choosing a career in the biological sciences…the knowledge that you leave the world a better place than you found it. But who gets to decide what is "better?"

Because the science of biology deals with the most precious of commodities—life—it is a discipline that is fraught with ethical dilemmas. In this chapter, we will examine three types of ethical problems that face the biologist: eugenics, laboratory animal research, and integrity in the gathering and reporting of research data. This is a very superficial treatment of the "gray area" of scientific ethics. Fortunately, many high schools and colleges have incorporated the study of research ethics in their science curricula. Case studies that describe controversial scenarios have been found to stimulate animated class discussions and help young scientists formulate their own personal ethical principles.

Everyday there are stories in the news about how advances in biomedical research improve, and complicate, our lives. Perhaps one of the most controversial areas of biomedical research is genetics. The same science that improves the breeding of livestock and crops to feed the world's burgeoning population strikes fear in the hearts of those who remember the atrocities committed in Nazi Germany in the name of racial purity. Eugenics, the study of hereditary control by genetic control, is a relatively new term that describes the ancient practice of selective breeding and the new technology of genetic engineering. For the biologist, genetic engineering is the key that opens a Pandora's box of good and harm. Do geneticists have the right to decide what is "normal" or "better" in cattle...and in people? As it becomes possible to eliminate human birth defects and diseases, will this advance humankind or is this just another form of the eugenics practiced in Germany? (Did you know that Canada and the United States had compulsory sterilization laws for the mentally retarded, "sexual perverts, drug fiends, drunkards, epileptics and diseased degenerate persons" that were not repealed until the 1970s?)

Perhaps more than anyone else, biologists understand and respect diversity in nature. With the input of geneticists, the Canadian National Reproductive Technologies Commission recently proposed guidelines for new laws that would set limits for the use of eugenics in humans. These guidelines would outlaw activities involving: sex selection for nonmedical reasons; research involving genetic alteration of zygotes; prenatal diagnosis for "susceptibility" genes; and the attempt to enhance human traits. With enactment of this legislation, geneticists and citizens hope that new biological technologies will be used for the good of continued individual autonomy while maintaining the collective rights of society at large.

Experimental research, especially in the biomedical sciences, has a certain glamorous appeal, and some discussions of careers in bioscience tend to equate science with experimentation. There are, however, other scientists who analyze and interpret an "experiment" that nature has performed. To presume that all biology is laboratory experimentation is as erroneous as the impression of some nineteenth-century philosophers that science is nothing more than observation and classification. The same qualities of enthusiasm, careful analysis, quest for truth, and open-mindedness are required, whether one's daily work involves laboratory experimentation, field observation, classroom teaching, routine laboratory testing, application of biology to quality control in industry, or administration and management of biological laboratories.

Those who perform biological experiments in the laboratory must of necessity make observations on living organisms. These organisms vary in complexity all the way from bacteria to human beings themselves. The elucidation of the functions of the human body has been made possible only by observations upon the bodies of animals most nearly resembling human beings; that is, the mammals. Many people misunderstand animal experimentation, and you no doubt have heard the accusations of those who depict the experimental biologist as cruel and sadistic, inflicting unnecessary pain and suffering upon the victims. These people have had no experience in experimental laboratories. After forty years of experience in such laboratories and acquaintance with many hundreds of workers in them, I can only say that the cruel and sadistic experimenter must be a rare bird indeed—I have never met one. I do know of cases, however, in which workers unable to learn careful and sympathetic handling of animals were dismissed from their jobs. There also have been cases of some experimenters criticizing others who in their opinion were not sufficiently careful to avoid pain.

Most experimenters know what their detractors fail to appreciate— namely, that observations made under conditions of extreme stress on the part of the animal subject cannot be considered to be made under normal conditions, and in most instances, the results might be invalid as answers to the questions under investigation. Experimental biologists have drafted a set of policies for humane treatment of animals in the laboratory, and a guide setting forth these policies has been assembled by the National Institutes of Health. It is a booklet entitled *Guide of the Care and Use of Laboratory Animals,* available for a small fee from the Superintendent of Documents, U.S. Government Printing Office, Washington, DC 20402.

If you are considering a career in biology, and if experimental biology appeals to you, you need have no fear that you will be in the company of unfeeling and inhumane monsters. By and large, scientists and other members of the so-called intellectual community lead in urging society toward a deeper understanding of all life and a lessening of cruelty, including an abhorrence of that cruelest of all activities, warfare. Moreover, of all people, the biologist must have a deep appreciation of all forms of life and a reverence for life itself.

In his paper "Integrity in Scientific Research," Mark Frankel, Director of the American Association for the Advancement of Science, described the current crisis in the ethical behavior of some scientists:

> Challenges to the quality and integrity of scientific research have become increasingly apparent in recent years with public revelations that some scientists have been guilty of fabricating data, falsifying results, and stealing the ideas and words of others. While few would contend that such egregious conduct is widespread, most scientists acknowledge that it warrants a serious and effective response. But while there is a consensus that fabrication, falsification and plagiarism are clearly unacceptable

behaviors, other types of conduct fall into a "gray area," where reasonable people may disagree over the proper course of action. These include, for example, practices related to selecting, sharing and reporting data and research results, allocating credit, the role of mentors, publications practices, peer review, whistle blower rights and responsibilities and the treatment of intellectual property

What contributes to questionable or unacceptable conduct of scientists? There is no single factor, but rather a complex set of intersecting influences that have combined to create a very stressful environment for researchers.... The complexity of scientific problems and advances in technology has also led to greater emphasis on large scale collaborative research projects. It is not unusual for several laboratories, perhaps located in different countries, to collaborate on particular studies. But the sheer magnitude of such collaborative studies can make it more difficult to guard against sloppy work and easier for less scrupulous researchers to contribute fudged data to the project. The desire to strengthen the competitive position of the United States in world markets has encouraged universities and government laboratories to join forces with industry in order to produce innovations of commercial value. Such partnerships bring benefits to all parties as well as to the nation's economic position in world markets. Yet, commercial relationships may pose potential conflicts with some of the traditional values of science, such as openness and prompt publication. These changes in the complexity, size and range of stakeholders that now characterize scientific research have placed demands on scientists that have outpaced the evolution of research norms and standards that researchers rely on as guideposts through rough ethical terrain. This has prompted the scientific community and others to reassess the usefulness of old standards and consider where new guidelines may be needed.

To promote integrity in research, the U.S. Department of Health and Human Service, the National Science Foundation, and other government agencies and research institutions have established policies regarding ethical conduct in research and have developed procedures to monitor research projects and investigate allegations of misconduct. Other research organizations, such as the American Association for the Advancement of Science have produced educational materials to prepare students to anticipate the ethical dilemmas they will face as research scientists and as responsible members of society who set the norms for acceptable moral behavior in science and other facets of life.

CHAPTER 3

WHERE WOULD YOU WORK AND WHAT WOULD YOU DO?

Before we explore the many varieties of career choices in the biological sciences, you might want think about where you would like to work and what type of working conditions you favor. For instance, you might already know that you would like to work with animals. Would you like to work outdoors in a zoo or is the research laboratory more your style? Or, if you think that you want to be a high school teacher, then planning your educational pathway will differ from that of a college professor. When you finish this chapter, if you already know where you would like to work and what you want to do, you might be more selective in your reading of Chapters 4, 5, or 6. And if you're still undecided about a career direction, then read them all.

EDUCATIONAL INSTITUTIONS

Educational institutions that employ biologists include secondary schools, both public and private; colleges; universities; professional schools; and many kinds of technical schools. Teaching can be one of the most rewarding professions you can find. One biology teacher in a secondary school has written that "the teaching of biology can be a dream come true, with all its attendant frustrations and rewards. I wouldn't trade it for anything."

In order to make science education more interesting for both students and teachers, the curricula in high schools and universities has become more "user friendly." The classroom and lab experiences are more integrated and emphasize problem solving and communication skills, not the acquisition of facts about biology. Teachers act as mentors and old didactic teaching techniques have been replaced with interactive learning, including allowing the students to select their own projects for lab exercises and classroom discussion. Even printed lab manuals have been replaced with computer software that allows students to select tutorials that resemble sophisticated PC games.

The employment outlook for teachers trained in biology is reasonably bright. Many new teachers are required to replace those retiring or leaving the profession. Many states are attempting to increase the science requirements for graduation and are having difficulty recruiting enough qualified teachers to implement their plans.

The 30,000 secondary schools in the United States constitute the largest market for teaching skills. The high school biology teacher is not charged with the responsibility of training professional biologists, but a good teacher will inspire some of the students to be interested in a biological career. The main task of the teacher is to educate responsible citizens, and in our increasingly technological society, a well-rounded person must have some knowledge of biological and other sciences.

The teacher in high school or junior college will usually be required to obtain a good background of biology courses in addition to studying teaching methods and science education. In general, teachers in four-year colleges and universities are not required to take courses in teaching methods, although no doubt some of them could benefit from such courses.

The biology teacher in secondary school is a professional teacher who has specialized in biology. Although the demands on the teacher's time are different from those of the research worker, he or she is equally challenged to keep up to date. A part of the job of teaching students to learn is to understand and apply the art of learning.

Not only is biology changing rapidly, but the methods of presenting biology to the students have undergone profound changes in recent years. The teacher must keep up with developments and is aided in doing so by such organizations as the American Association for the Advancement of Science, the American Institute of Biological Sciences, The National Science Teachers Association, and the National Association of Biology Teachers. These organizations publish periodicals and hold meetings that inform the teacher of new discoveries, as well as new ideas in teaching. Summer courses are available at many colleges and universities and at biological stations, which are usually operated by universities. Teachers in secondary school face a special challenge in the great diversity among their students, some of whom may never take another science course, while others will exhibit unrestrained enthusiasm for the subject and will go on to a full-time career.

Although most biology teachers are specially trained, some have entered the profession from other disciplines simply because they find biology more interesting than their former employment. These recruits sometimes are teachers from other subjects and sometimes are biologists who have worked in industrial or government laboratories or elsewhere and have decided that they prefer teaching. The college training for biology teachers in secondary schools includes a broad study of many biological subjects, including both laboratory experience and fieldwork and both plant and animal material. Courses in chemistry and physics are also essential; indeed, in some small schools, a teacher may be called upon to teach more than one science. The college student intending to teach biology should gain as much experience as possible in practice teaching. He or she also will find it advantageous to obtain summer employment in a biologically oriented job, such as work in medical or research laboratories, conservation projects, or forests.

Most biology teachers in elementary or secondary school have entered the profession immediately after completing college, but more and more find it desirable to continue study until a master's degree is obtained. The standards for high school teachers are being elevated, and in some of the better school systems, the master's degree is the ticket for entrance. Training programs covering a five-year period and culminating in the master's degree are available in some training centers for teachers. The Ph.D. or Sc.D. degree is primarily a research degree. Although generally needed for careers in teaching in a university, it is not required for secondary school. This is just as well, for the emphasis on research that is necessary during the study for the doctorate is of limited value for the kind of work the high school teacher will be doing.

High school teachers are highly respected members of the community, and in recent years their prestige has grown. There are many rewards, both tangible and intangible, for devoted teachers, and working conditions and pay scales are improving constantly. There is never an oversupply of competent and devoted teachers.

As employers, high schools generally use different criteria in hiring than do other employers of biologists. The high school teacher will usually be required to have studied courses in teaching methods and in science education in addition to biology and related sciences. The beginning teacher will be expected to have a bachelor's degree, or preferably a master's degree. In addition, such teachers may be expected to continue their education by summer study or other means.

Unlike the secondary teacher, the university teacher of biology is likely to consider herself or himself as primarily a professional biologist and only secondarily a teacher. College teachers usually earn a doctoral degree before becoming full-time faculty members, and doctoral training is strongly slanted toward stimulating interest in research. The advanced degree, Ph.D. or Sc.D., is awarded as a result of concentrated study of a highly specialized field, and colleges will consider for hiring only those individuals whose specialized study is closely related to the subject that the new instructor is expected to teach. If you obtain a Ph.D. in a department of education, for example, you cannot expect to obtain a job teaching in a university biology department.

The preparation for a teaching career in a college is the same as that for a full-time research worker. Most colleges and universities, especially the larger ones and those with the greatest prestige, expect faculty members to combine teaching and research careers. In that way, the college teacher not only passes along to the next generation the knowledge accumulated in the past, but also keeps at the forefront of new developments. The young man or woman just beginning a teaching career in college may be given the title of instructor, but with some experience may look forward to promotions—to assistant professor, then associate professor, and finally professor. With each increase in rank there is an increase in salary, greater opportunities to teach more advanced courses and guide young graduate students who will be the next generation of college teachers, and greater responsibilities in serving on committees that help guide the policies of the institution.

Colleges offer many attractions for those considering a teaching career. They are frequently located in pleasant towns or in cities with many cultural advantages. The faculty members form a community of people with a similar level of education and with many interests, hopes, and aspirations in common. To an extraordinary degree, the teacher is her or his own boss, and one is free to plan one's own time and to do the work in one's own way.

Many universities have several departments that employ biological scientists while smaller colleges may have but a single department of biology. Larger ones may have separate departments of botany and zoology. Many other arrangements are possible, and there may be several departments representing the various biological disciplines. Colleges of agriculture, with their emphasis on applied biology, utilize an especially large variety of biological skills. Specialists in several biosciences are to be found in colleges of medicine, dentistry, pharmacy, and veterinary medicine; these schools are usually integral parts of universities, but some are independent.

It might be helpful for you to visit one or more of the colleges or professional schools near you, seek out the biologists working in them, and talk with them. The more kinds of biologist you talk with, the better your own judgment may be about selecting a similar kind of career. You will find most of them glad for a chance to talk about their work and especially happy to welcome a prospective recruit to their profession. But don't expect them to make up your mind for you; you must do that yourself.

If you are considering biology teaching as a career, you should not overlook the possibilities offered by junior or community colleges. Although the courses of instruction are not as advanced as in four-year colleges, most of them offer courses of sufficient diversity to interest most biologists. In some instances, their programs of general biology are at least as good as those of the lower divisions of four-year colleges. Unlike the four-year college, the junior college often requires applicants for teaching jobs to obtain teaching certificates, but requirements for certification of junior college teachers differ from state to state. For specific information, contact your state department of education.

The teaching load in junior colleges—the number of hours per week actually spent in the classroom—is usually somewhat less than that required in high schools, but greater than that demanded in senior colleges. The senior college teacher is expected to spend a significant portion of his or her time on scholarly pursuits such as writing or research. Often, salary scales are lower in junior colleges than in either high schools or four-year colleges, but in recent years, they have improved. The teacher who wishes to spend part of the time in research will find that the junior college offers less opportunity than does the senior college. There are, then, some frustrations accompanying junior college teaching, but the dedicated and enthusiastic teacher finds that they add to the interest and challenge of the job. Some teachers feel that the junior college offers the most exciting and rewarding field in the educational world.

GOVERNMENTAL AGENCIES

The largest employer of biologists or of those whose work is closely related to biology is the federal government. Biologists may be found in most of the departments. Let us look, for example, at the Department of Interior, which is charged with the duty of conservation of natural resources. A list of the agencies within that department that employ bioscientists and biologically oriented people includes the Bureau of Land Management, National Park Service, Geological Survey, Bureau of Reclamation, Fish and Wildlife Service, and Bureau of Indian Affairs. The National Park Service, for example, has about a hundred natural areas where there are employees with a background in the life sciences. The majority of the national park superintendents have training in bioscience together with many years of experience within the park system. Life science majors begin as park rangers; the competition for the positions is keen. Rangers have to deal not only with environmental problems but with people problems as well.

Even the large number of positions for bioscientists in the Department of Interior is exceeded by the Department of Agriculture, which is the largest employer of biologists.

The multiplicity of jobs within the federal government is indicated by the following list—a very incomplete list—issued by various agencies and classified in biological sciences:

Agricultural bacteriology	Forestry research
Agricultural technology	Gardening
Animal physiology	Genetics
Bacteriology	Herbarium aide
Biological aide	Horticulture
Biology	Hydrology
Cereal technology	Illustrator
Cotton technology	Investigator
Dairy husbandry	Librarian
Ecology	Medical biology technology
Editor	Microanalysis
Entomology	Microbiology
Fish culture	Mycology
Fishery research biology	Nematology
Forestry	Nuritionist

Parasitology	Range management and
Park naturalist	conservation
Park rangery	Seed technology
Pharmacology	Soil science
Plant pathology	Systematic zoology
Plant physiology	Tree culture
Plant taxonomy	Wildlife management
Poultry husbandry	Wildlife research biology
Predator and rodent control	Zoology

The above list of forty-eight jobs in various government agencies is a purely arbitrary one of actual titles of positions selected by the author and arranged alphabetically. You can judge its incompleteness by the fact that the Department of Agriculture alone has a list more than twice as long. Some of the categories, though, seem to be less directly related to biology than are those above.

The Internet websites of many government agencies contain career information including a list of current openings and application procedures. See the appendixes at the end of this book for more information. The U.S. Fish and Wildlife Service website is a good example.

A career in government service has many attractive features for biologists. Employment conditions are good, and the increasing emphasis on life sciences bodes well for the future. The levels of jobs range all the way from the blue-collar jobs for those not yet fully trained to the so-called supergrades for distinguished scientists.

For those who start to work before completing their training, the government offers educational assistance programs. Arrangements are made for the employee to receive full pay while attending classes at a nearby university part-time, or even full-time up to one year. After completing this training, of course, the employee is in line for a job with a higher rating.

Government positions offer a salary scale close to that found in private industry and somewhat better than that in the educational world. The government scientist has good laboratories and equipment and much freedom in doing the job her or his own way. The scientist in government can publish the results of the work in scientific journals and hence achieve a reputation in the world of science. Part or all of the expenses for attendance at scientific meetings also will be reimbursed.

Other fringe benefits, such as group life insurance, health insurance, and retirement pay, are comparable to those available elsewhere. A scientist who works in the laboratory or in the field may wish to continue to do so until retirement or may have an opportunity to become an executive and help administer the programs of the department. Either way, there are opportunities for advancement.

You can obtain information about government jobs by contacting the Federal Job Information Center; there are more than a hundred such centers across the country that provide local job information. They are listed under "U.S. Government" in the telephone directories of metropolitan centers, or you can obtain the local number by dialing 800-555-1212 (the telephone information number).

Although the government service is centered around Washington, DC, employees of such agencies as the Department of Agriculture, the Department of the Interior, the Public Health Service, and others are to be found in all the states, and in some instances, their duties take them overseas.

We should not leave the areas of federal service without mentioning The Food and Drug Administration (FDA). This important agency, a part of the Department of Health and Human Services, employs a large number of biological scientists and has a continuing need for biologists, chemists, and other scientists of the highest caliber. The FDA is charged with the duty of ensuring that foods, drugs, and cosmetics sold in the United States are safe, properly manufactured, honestly advertised, and effective for the purpose for which they are sold (some foods, especially meats, are inspected by the Department of Agriculture).

Each of the states, and some cities, employ biological scientists in roles similar to many of the federal jobs. Biologically oriented agencies of state and local governments include fish and game commissions, parks, aquariums, arboretums, and museums. On the average, these positions pay slightly less than comparable jobs in the federal service, and such fringe benefits as vacation time, sick leave, insurance, and retirement pay are not quite as generous as those enjoyed by federal workers.

In summary, whatever your interest in biology might be—whether you would like to work outdoors in field or forest, streams or seas, or indoors in a laboratory or at a desk, in routine work or in research, in doing your own work or in supervising and administering the work of others—a place can be found in government service. If, however, you want to experience the thrill of bringing young people into their first contact with the exciting world of biology and watch them mature into scientists in their own right, you will do better in teaching.

INDUSTRY

Most of the research and development in our country is performed in industry, but there are a great many biologically oriented jobs outside of the research function. For example, many industries depend upon microorganisms for production of their products. Some of these, such as cheeses, alcoholic beverages, and baked goods, have a long history, with records going back to about 6000 B.C. Others are of recent origin and owe their existence to new developments in biological science; prominent among these is the manufacture of pharmaceutical products by enzymatic action of microorganisms in large vats, followed by extraction and chemical isolation of the desired products. Some industrial biologists are using microorganisms for the production of methane and alcohol from waste products as alternate energy sources. Even the yields of minerals in low-grade ore deposits are being improved by microorganisms in some processes. Purification and recycling of water also gives employment to some biologists in industry.

Individuals with biological qualifications are employed by an astonishing variety of business and industrial concerns. An indication of the importance of this source of employment for biologists can be obtained by listing some of the industries involved. The chemical industry offers many opportunities. Prominent in this category are those firms producing agricultural chemicals, including substances used for controlling pests or stimulating crop production, as well as dietary supplements for the nutrition of animals and human beings.

Among those offering positions with biological orientation are processors of foods and beverages, the manufacturers of cosmetics, the agricultural industries needing the skills of those trained in animal or plant husbandry, the breeders of fur-bearing animals, fisheries, forest products companies, and even such concerns as the manufacturers of textiles, leather goods, and petroleum products as well as public utility companies and the aerospace industry. Still others are the publishers of biological books, manufacturers of laboratory equipment, general laboratory supply houses, and those companies that make a business of collecting, culturing and processing biological material for the use of educational institutions and research laboratories. There is scarcely any biological discipline mentioned in this book that is not represented in one or more of the industries just listed.

One growing industry is made up of companies that perform scientific services under contract for others. Some of them do biological testing exclusively, while others include biology but have other departments as well. The customers of the biological testing companies include the U.S. government, drug companies, cosmetic manufacturers, and a variety of other individuals and corporations. Many of the customers have testing laboratories of their own but sometimes need the additional help of the service companies. Although much of the work is routine, the testing laboratories also do a considerable amount of research, and some of their bioscientists are highly regarded by their colleagues in other organizations. The biological disciplines represented in this industry include nutrition, biochemistry, physiology, pharmacology, microbiology, cytology, histology, toxicology, and pathology.

Manufacturers of pharmaceutical products probably employ more biologists than any other industry. The range of jobs in the drug industry includes biological manufacturing and packaging; quality control; biological testing of products; writing of brochures, reports, and correspondence; and training of workers. Especially noteworthy is the magnitude of the research efforts. More than twenty thousand people are employed in the research laboratories of this one industry, and an extraordinarily high percentage of them have a background in biological science. The demand for biologists and other scientists in the pharmaceutical industry continues to grow.

For our present discussion, the significance of such figures is that career opportunities are offered for those with the training to take advantage of them. The scientist in the pharmaceutical industry is, by and large, given whatever is needed in the way of materials, equipment, and assistants to do the job. The work is expensive, partly because of some 175,000 new chemicals tested in a year, not more than 8 out of each 1,000 show enough promise to warrant testing in human beings, and in some instances, all but 1 in 15,000 are rejected before they reach the market.

In spite of these difficulties, the industry points with pride to the fact that over the past thirty-five years 90 percent of all the world's drugs—nearly six hundred new chemical entities—have been discovered and developed by scientists in its employ. Most of these scientists are either chemists or biologists, and their discoveries have revolutionized the treatment and prevention of disease.

The research biologist in the drug industry is able, to a degree unmatched anywhere else, to work in collaboration with scientists in disciplines other than his or her own, whether it be in physical sciences or other biosciences. Cooperation between biologists and chemists is a daily occurrence, and the biologist often consults with physicists, mathematicians, psychologists, and pharmacists. Bioscience disciplines represented in the industry include physiology, biochemistry, toxicology, pharmacology, pathology, animal husbandry, microbiology, immunology, systematic botany, entomology, and nutrition.

Papers and monographs written by industrial scientists are published in the leading scientific journals and books, and research workers in industry often hold high positions in scientific societies. These are developments of the past few years, for many of us can remember when scientists who worked for industry—especially biological scientists—were held in low esteem by their academic colleagues. A major reason for improvement in this attitude is the enlightened attitude of many companies that have come to realize that scientists require a different kind of environment than do people with commercial interests. Now, there is a fair amount of interchange between the academic world and industry; some university professors forsake academia for jobs in industry, while many scientists leave industry for teaching positions. In addition, many industrial scientists hold part-time teaching appointments in nearby universities.

The professional levels of biologists in the drug industry range all the way from beginners with baccalaureate degrees working as technicians to scientists of international renown. The beginner is encouraged to continue his or her education while receiving full pay from the company; most companies pay the tuition charges for part-time study at a nearby college or university. Many of the pharmaceutical laboratories are built in pleasant rural or suburban locations, in keeping with efforts to make working conditions as pleasant as possible. Salaries are slightly higher than in colleges, and the fringe benefits of vacations, sick leave, insurance, and pensions are among the most liberal of any industry.

Jobs with biological orientation but outside research departments are becoming more common in industry. Some companies prefer professionally trained representatives to present their products to customers. These are not salespeople in the old sense but individuals who provide a liaison between company and customer. Others employ professional people in training programs for sales personnel and technical representatives. If you can combine technical competence with a flair for writing, you will find that there is a brisk demand for staff writers and editors. In summary, despite obvious differences between the academic life and industry, you will find the same types of people in both, and there are as many varieties of opportunities in one as in the other.

If you are interested in a career in industry, try the Peterson's "virtual" education and career center website (www.petersons.com). You can search for biology programs in undergraduate and graduate institutions and develop a list of potential corporate employers in the area of interest by using the Peterson's search site.

SELF-EMPLOYMENT OPPORTUNITIES

One other commercial outlet for biological talent remains to be discussed: namely biologists who go into business for themselves. We have already discussed the biological testing industry; several of the laboratories in that category were started by biological scientists. Starting such a project requires an amount of capital that is not readily available to everyone. Easier to start are laboratories that perform testing services for physicians and hospitals. Although most hospitals have their own laboratories for biological and biochemical examinations, there also are large numbers of independent laboratories performing these services. To be successful in such a venture, you should have a thorough background in the technical aspects of biochemistry, hematology, and the preparation of microscope slides. The more successful of such ventures have proven to be highly rewarding financially.

Self-employment opportunities are also available for botanists. Some of them become greenhouse operators who may engage in a particularly lucrative specialty, such as the growing of orchids. Botanical systematists become consultants to industry. Plant pathologists are also in demand as consultants, and some of them go into private practice. The number of self-employed botanists is small, but some find that their business pays well.

The collection, preservation, and sale of biological specimens is another source of self-employment. Vast numbers of living and preserved frogs, turtles, earthworms, sharks, sea urchins, anemones, seaweeds, seed-bearing plants, and other plant and animal organisms are used for teaching and research in schools, colleges, and research laboratories. The demand for them in the past has been met almost entirely by collecting wild specimens, but with the growing threat to the survival of many species, specialized farming must be considered as a source for the future. The demand fluctuates with the number of students of biology and the generosity of school budgets for materials.

For the senior scientist, independent consulting has become a lucrative occupation. Law firms are always looking for experts who can evaluate and corroborate evidence to strengthen their cases and, in fact, there is an entire field called "forensic biology." The growing interest in the preservation of our natural resources has nurtured opportunities for consultants in almost every area of ecology. Biologists act as experts who prepare "impact studies" for new construction projects or can consult with lawyers who counsel industries whose work affects the environment. Consulting biologists also may be called up to draft legislation or serve on governmental or organizational committees that are charged with developing policies dealing with concerns such as ethics.

INDEPENDENT RESEARCH LABORATORIES

Research laboratories that are neither directly connected with industry nor a part of a specific department of a university receive their support in a variety of ways. Some of them depend largely upon the income from endowment funds that have been donated by an individual or group wishing to use their wealth in a constructive way. Others depend more upon annual contributions. Most of them accept grants of money from government funds or from interested industries. Whatever their sources of income, large numbers of bioscientists ranging in rank and status all the way from laboratory assistants to Nobel prize–winning scientists find a stable and congenial atmosphere for their research. The same wide range of biological and biomedical disciplines is to be found in the independent research laboratories as in universities or industrial installations. The qualifications for appointment are the same, and salaries and fringe benefits do not differ greatly. In some of them, it is also possible to participate in the education and training of young scientists and to hold faculty rank in a nearby university.

There are independent laboratories in every section of the country. The following few examples will illustrate the wide variety of interests among these institutions, and you may note that some of them are so well known as to be almost household words:

- Jackson Laboratory, Bar Harbor, Maine 04609, is famed for genetic studies, especially on unique strains of mice.
- Marine Biological Laboratory, Woods Hole, Massachusetts 02543, is a favorite spot for many biologists in the summertime, but it is active the year-round in collecting and culturing marine organisms and in biological research.
- Wistar Institute of Anatomy and Biology, Philadelphia, Pennsylvania 19104, is not only a research and teaching center, but has been known as a source of laboratory animals, such as the famed "Wistar Rat."
- Worcester Foundation for Experimental Biology, Shrewsbury, Massachusetts 01545, is known for work on endocrine physiology, especially for its origin of birth control pills, but it carries on other biological work, too.
- Mayo Clinic and Foundation, Rochester, Minnesota 55905, is a hospital as well as a laboratory for a wide variety of research and training in the biomedical sciences.
- Salk Institute of Biological Studies, San Diego, California 92138, carries on advanced research on cancer and in immunology and other disciplines.
- Midwest Research Institute, Kansas City, Missouri 64110, concentrates on general biological sciences, health sciences, and environmental science.
- Southern Research Institute, Birmingham, Alabama 35255, engages in research in biology, chemistry, and various other sciences.
- World Life Research Institute, Colton, California 92324, focuses on such problems as marine biotoxicology and environmental pollution.

BOTANICAL GARDENS AND ARBORETUMS

If you have never visited a botanical garden or arboretum, it may surprise you to learn that there are many kinds of careers available in these interesting places and a considerable demand for men and women to fill staff positions. Botanical gardens are more than just beautiful displays of trees and flowers. They typically have many varieties of plants growing in the open and more in greenhouses and conservatories. The care of these plants is much more demanding than that of the usual commercial greenhouse or nursery. The latter will have no more than a few dozen kinds of plants, but a large botanical garden may have several thousand different species of plants, each with its own particular requirements for growth and propagation. In addition, botanical gardens and arboretums maintain collections of dried specimens. These collections can be very large—the New York Botanical Garden has more than three million specimens.

These gardens are not merely repositories of stationary specimens. They are also educational institutions, and many of them also conduct extensive research. Some of the subjects of research are ecology, plant anatomy, systematic botany, economic botany, plant physiology, phytopathology, and biochemistry. Sometimes research botanists and curators hold professorships in nearby universities. Graduate students in some departments of botany can complete all their research for advanced degrees at a botanical garden or arboretum.

Botanical gardens offer educational programs to schoolchildren and to the general public. These programs consist not only of labeled collections of plants, but also of exhibits, classes, lectures, publications, and the furnishing of background material in botany for schools of all levels. This is a two-way street, with students being given an opportunity to visit the botanical garden and with the staff enjoying contact with teachers and students in the classroom. Staff members especially skilled in these activities have not always been easy to find, and directors of the gardens are on the lookout for promising candidates.

The kinds of jobs available in botanical gardens and arboretums include gardeners, horticulturists, caretakers for the herbarium, staff members for preparing exhibits, directors of educational activities, editors and writers for the publication programs, librarians, and research scientists. In some instances, there are opportunities for travel to remote parts of the globe for the systematic exploration and collection of new (or old) plants. This may serve many purposes, such as adding to the collections, improving the exhibits, aiding in the preservation of plant species threatened with extinction, and discovering new species of value as ornamentals, as food plants, or as sources of new drugs.

If you wish to consider a career in this field, you should visit at least one botanical garden or arboretum, and preferably more than one, since they differ markedly in size, organization, and activities. Make an appointment with the director or other staff member, discuss your interests with him or her, and observe the place in action.

ZOOS AND AQUARIUMS

What the botanical gardens do for plants, zoos do for animals and aquariums do for marine life. In addition, since animals have such fascination for people of all ages, attendance at zoos and aquariums exceeds that at botanical gardens. In fact, no professional sport draws as many spectators as do zoos or aquariums. They provide public recreation and outdoor amusement, but more importantly, they are educational and research institutions. The educational activities of zoos and aquariums include informative exhibits for the general public, the publication of guidebooks, tours for schoolchildren, orientation programs for school-teachers (sometimes courses are given in cooperation with local colleges so that teachers may earn credit), public lectures and films, television programs, and facilities for the use of advanced students in the study of animal behavior. There are sometimes special programs for particular groups such as clubs interested in birds or some other animal groups, 4-H Clubs, Boy or Girl Scouts, photographers, artists, or others.

The importance of these activities varies with the size and resources of the zoo or aquarium. Most are carried on at the facility, but there may be classroom follow-ups with specimens taken to schools. The employees most often encountered by the public are the groundskeepers, zookeepers, and their assistants. Keepers engage in practical animal husbandry; recently there has been a tendency to elevate the positions professionally and to demand more training than in the past. Curators are the professional zoologists in the zoo or marine biologists at the aquarium, and positions as curators are often held by individuals with master's or Ph.D. degrees. Curators supervise the care of the animals and make decisions on the choice of animals to display and the means used to present the displays to the public.

The zoologist or marine biologist may perform research in animal breeding, genetics, ecology, animal behavior, or other topics. He or she may need to travel to visit other facilities, to study animals in the native state or to collect specimens. Although the most familiar animals in the zoo may be the mammals and fish at aquariums, there also must be experts on birds (ornithologists), reptiles and amphibians (herpetologists), and fish (ichthyologists). An aquarium for the study and display of fish and other aquatic creatures is often associated with a zoo, but it may be separate. Workers at zoos and aquariums find that their work never gets dull, they are relatively free from bureaucratic regulations, and their salaries are comparable with those of people of corresponding education and responsibility in local schools and colleges.

MUSEUMS OF NATURAL HISTORY

Most of us like to make collections of interesting objects, and the greatest collections of objects of scientific interest are in natural history museums. Museums play an important role in transmitting scientific information to the public; hence, they are of increasing significance as informal educational institutions. The most visible evidence of the museum's educational function is the exhibit. Museum exhibition as a profession has a strong appeal to certain persons, but the employment opportunities are limited. Often exhibits are prepared by staff members whose primary duties lie elsewhere, but there is an increasing tendency for museum exhibition to become a profession in its own right.

Exhibits are educational displays. Their preparation involves conception and planning followed by design and execution. The exhibitor, then, is less of a specialist than are most biologists. Not only must the person be well grounded in general biology, but he or she should especially understand systematics, ecology, and conservation. In addition, it is important to know something about the principles of three-dimensional design, color harmony, and photography. The exhibit designer will have some contact with anthropology, paleontology, geology, and geography—he or she does not need to be an expert in these fields but should have sufficient understanding of them to be able to collaborate with professionals in such disciplines.

The exhibit is by no means the only educational medium of the natural history museum. Many museums contain classrooms, and museum personnel often collaborate with local school systems in informal instruction either in the museum or in the schools. The museum's collections are also important sources of material for instructional purposes or for scholarly research. The curator in a large natural history museum is generally a systematic biologist and has duties that are manifold: He or she occasionally makes field trips to remote areas for specimens. Once the specimen is in the museum's collection, it has to be properly identified, documented, labeled, and maintained in condition for proper use by scholars not only now but in future generations. The curator also helps in the exhibit programs, collaborates with colleagues in other institutions, gives lectures—both technical and popular—and must do scholarly research in the field and publish the results in technical journals. In many instances, the curator holds an appointment on a university faculty (some important museums are integral parts of universities) and aids in the instruction of advanced students, especially graduate students.

The organization of the staff of a large museum, such as the National Museum of Natural History, a part of the Smithsonian Institution in Washington, DC, resembles that of a university. The areas of interest to biologists are divided into departments, such as the Department of Vertebrate Zoology, Department of Invertebrate Zoology, Department of Entomology, Department of Botany, and Department of Paleobiology. Within each department are divisions staffed by biologists of various ranks such as curators, associate curators, and assistants, reminiscent to the academic ranks of professor, associate professor, and assistant professor. Most museum directors regard their curatorial and exhibit staffs to be undermanned and seek to recruit young trained personnel within the limits of their budgets. Museums have caught the popular imagination and are increasingly being supported by the general public. Much of this support, however, is in the form of volunteer unpaid work. Many of these volunteers are well-educated retirees or others not needing full-time paid jobs. In museums they work as docents (guides and lecturers), laboratory assistants, and in the field. Although the activities of these volunteers reduce the necessity for hiring full-time employees, the increased interest in museums as exemplified by them helps to draw additional funds available for recruiting young people for full-time employment.

Positions as assistants in natural history museums are available in limited numbers. They are generally filled by applicants who have bachelor's degrees with majors in biology. In former years, many museum assistants were eventually promoted to curatorial positions, even without additional formal education. With the increasing emphasis upon research in systematic biology, appointment or promotion to the position of curator today goes only to those who have completed the training for the Ph.D. degree. Many museums have training programs for technicians, and the technical assistant may be able to obtain the formal education necessary for promotion.

The jobs available in museums do not compare in number with those in teaching, industry, or government research, but for the limited number who find employment in them, museums offer an enjoyable environment. There are few careers that instill greater dedication on the part of their devotees. If you are interested in pursuing further the possibilities of a career in museum work, you should visit a museum near your home and talk with a curator, or you may write to one of the major museums, such as the American Museum of Natural History, Central Park West at Seventy-ninth Street, New York, NY 10024, or the National Museum of Natural History, Tenth Street and Constitution Avenue NW, Washington, DC 20008.

THE PEACE CORPS

We should not leave the subject of nonprofit employers of biologists without a word about serving as a volunteer in areas where trained people are needed. You may not wish to serve in volunteer organizations as a permanent occupation, but a year or two spent in helping others may give valuable experience in addition to the satisfaction that comes from a useful job well done. There are many groups offering opportunities to be of service, and most of them do not specifically call for people with biological training. An exception is the Peace Corps, which has had many requests that can best be satisfied by individuals with biological knowledge. Peace Corps volunteers receive no salary, but transportation and health needs are cared for, and monthly allowances cover modest living costs.

Biologically trained volunteers have served in many parts of the globe. Most of them have taught in secondary schools, in teacher training schools, or they have assisted in fieldwork. Included among the fields of biology for which there have been calls in the past are the following:

• Training teachers in general biology, botany, bacteriology
• Medical technologists to train hospital technicians
• Help in freshwater fisheries and fish farming

- Assistance in livestock breeding
- Entomologists to aid in pest control
- Plant pathologists for help in controlling plant diseases

Information can be obtained by writing to Peace Corps, 806 Connecticut Avenue NW, Washington, DC 20526.

CHAPTER 4

THE MANY VARIETIES OF BIOLOGICAL CAREERS

The group of sciences known as the biological sciences is so large and varied that there are many possible ways of classifying the numerous types of careers available. For example, they might be grouped according to subject matter: whether plant or animal or microorganism; whether the object of attention is a whole organism or the cells of which it is composed; or whether one's interests are primarily in research or teaching or in practical applications. There are educational careers, industrial careers, health and allied professions—these and many more will be considered in the discussions that follow.

Growth and changes in the biosciences occur at such a breathtaking pace that it is difficult to keep up with them, and some of the most promising careers are in areas that were scarcely known only a few years ago. Frederic Joliot-Curie, one of the great physicists of the early twentieth century, reportedly said that while the first half of this century belonged to the physicist, the second half would belong to the biologist. Some of the recent developments in biology seem to bear out this prophecy, and perhaps we can do no better in opening our discussion of various fields of bioscience than to start with some of the newer types of careers.

BIOTECHNOLOGY

Biotechnology has been characterized as a "wonder world," which promises to revolutionize the ways in which we live, work, and play. Rapid developments in biotechnology are the payoff of the cracking of the genetic code stemming from the discoveries of the structures of genes and DNA. The potential of this research has been compared by *U.S. News & World Report* with "the discovery of fire, invention of the printing press, and the splitting of the atom." Whether this is an exaggeration or not, there is no doubt that the accomplishments already are impressive, and the future promise can scarcely be imagined. High school biology students today can perform experiments that the most sophisticated laboratories could not attempt a short while ago, and new jobs and even new kinds of careers are opening up.

Biotechnology uses cells, the building blocks of life, to create new or improved products such as pharmaceuticals, diagnostic technologies, and agricultural and environmental goods and materials.

Examples of what has already been accomplished are not hard to find. Members of the older generation can remember when diabetes was a certainly fatal disease. Then, in the 1920s, it was discovered that an extract of the pancreas of animals could aid in the treatment, and frequently prolong the lives, of diabetics. There were drawbacks, however. The animal insulin was not quite chemically identical to human insulin, and allergic reactions were encountered occasionally. Furthermore, it required a great and expensive effort to supply the demand. This state of affairs lasted for more than half a century, and by the year 1981 the pancreas glands from some 23,500 pigs and cattle had to be obtained from slaughterhouses for processing. Then came the biotech revolution. Using gene splicing techniques, appropriate fragments of human DNA, capable of catalyzing the production of genuine human insulin, could be inserted into a microorganism. The resulting bacterium—which can be considered a new form of life—needed only to be placed in the proper culture medium for economical production of the first pharmaceutical product to be marketed as a practical result of this research.

Other pharmaceutical products from similar research efforts include the protein tissue plasminogen activator (tPA), which can reduce the damage of heart attacks by dissolving blood clots, and a variety of vaccines against such diseases as malaria, rabies, hepatitis, venereal diseases, foot-and-mouth disease in animals—and even cancer. One much publicized substance is interferon, a protein produced in our cells when exposed to a virus. Nothing has given greater hope as a "wonder drug" against an assortment of ailments ranging from the common cold to cancer to AIDS. So rapid are advances in research on this fascinating material that anything written here will already be out of date by the time it is in print—a complaint often expressed by workers in biotechnology.

I would not want to leave you with the impression that drug manufacture is the only use of recombinant DNA techniques. Equally exciting, and perhaps of even more importance in the long run, are projects in the agricultural realm—food crops that resist the assaults of insects, or that manufacture some of their own fertilizer (as peas and beans do now), or that will grow in soils and climates that cannot now produce food. Whole new species of plants have been created by gene splicing, such as crossing a potato and a tomato, and even a bean with a sunflower.

Biotechnology is not exclusively concerned with gene splicing but includes several other types of techniques. For example, the technology of cell fusion, unlike DNA manipulation, deals with whole biological systems rather than small fragments. Cells that are formed by fusion of two different types of cells form a special type of new organism called a hybridoma. Hybridomas are capable of producing antibodies against a wide variety of diseases, and they are already being used as tools in diagnosis of many kinds, including bacterial diseases, prenatal disorders, and cancer. An even greater potential may be foreseen in the use in therapy of the antibodies—called *monoclonal antibodies*—produced by hybridomas.

Other types of biotechnology include test-tube breeding of farm animals with embryo transfer, so that ordinary cows can give birth to super cows, for example, which can produce as much milk as three common cows. Many thousands of such superior offspring are already being produced. Genetic engineering also may produce superior meat animals. One could go on and on, limited only by the fertility of one's imagination. Some day there may be improved microorganisms for turning waste into a useful energy source, microbes that increase the yields of oil wells or that digest oil spills, sulfer-metabolizing organisms that take care of important sources of air pollution—perhaps even bacteria that can manufacture proteins that will be far superior to silicon and metal as computer chips.

What does all this have to do with your choice of a career? Partly this: There has been an explosion of numbers of companies entering this field, and such firms are scrambling for geneticists, molecular biologists, and other specialists. One marketing research firm estimated that there would be a high demand well into the twenty-first century. Another research and development firm has been quoted as saying that a person in this field—termed bioengineering—after four or five years of experience "can command pretty much what he or she wants." It should be understood that landing such a job and commanding a high salary require thorough preparation. Part of this preparation would include several years of study beyond college, with a doctorate in a biological science such as genetics, microbiology, or molecular biology. This is a formidable and challenging prospect, but it should be remembered that for each job at such a level there would be a demand for supporting personnel. People with a baccalaureate or master's degree are also in demand at slightly lower salaries. Those who start at these levels can look forward to opportunities for further training on the job with appropriate advancement or to obtaining credits toward a doctorate by studying while working. Some may wish to pursue such a course, while others will find a rewarding and satisfying career in the supportive role.

GENETICS

In the above discussion of biotechnology, it is clear that application of the science of genetics offers many opportunities to make profound changes in the world in which we live. Indeed, a discussion of genetics must be in large part a continuation of the commentary on biotechnology, although there are aspects of genetics other than genetic engineering.

The study of genetics pervades almost every field of biology today and is ever present in the news. Whether it is the cloning of a live sheep or a breakthough in a cure for AIDS, genetic research will be on the cutting edge of the biological sciences well into the twenty-first century.

Simply defined, genetics is the science of heredity. Its pursuit involves observation of the superficial appearance of an animal or plant as related to its parents or its offspring, the inheritance of biochemical characteristics, the intimate nature of the structures in cells—especially the sex cells—that transmit heritable traits, and all other aspects of the complex mechanisms involved in heredity. The work that individual geneticists do varies from the solution of highly abstract and theoretical problems to the application of genetic principles to practical and economically important subjects. Any species of plant, animal, or microorganism may serve as the object for the geneticist's scrutiny. Much of what is known about cellular mechanisms involved in the transmission of traits from one generation to another has been gathered from the study of relatively simple forms of life, such as bacteria and yeasts. The lowly fruit fly, *Drosophila,* has probably contributed more than any other animal to knowlege of the cellular structures known as chromosomes and genes, which are involved in heredity. Among higher plants, probably more is known about the genetics of corn than of any other plant.

Human genetics has been widely studied, too, including hereditary traits, hereditary diseases, and the interrelationship between inherited characteristics and their modification by the environment. There are, however, many gaps in our knowledge. Much has been learned from the study of identical twins and their comparison with nonidentical (also known as fraternal) twins. Since identical twins develop from a single egg fertilized by a single sperm cell, they have the same genetic makeup. Nonidentical twins, on the other hand, are no more alike genetically than are ordinary brothers and sisters, since they develop from two eggs fertilized by two sperms.

Genetics has conferred many varied practical benefits upon humankind. The great hybrid corn industry, which has added millions of bushels to the annual crop of that important food, owes its existence to the study of the genetics of the corn plant. Developments in the genetics of crops, especially rice, have contributed greatly to the widely heralded *green revolution* that has increased the production of the staple food so important to the millions of people in Southeast Asia. Perhaps further advances will lessen the dependence of these new varieties on the use of massive amounts of fertilizer and pesticides.

The selective breeding that has produced superior strains of livestock is a practical application of genetics. Several years ago a parasitic fly, the screw-worm fly, threatened to wipe out the cattle industry in Florida; the larva of this fly feeds upon the living tissues of cattle. Geneticists (and other biologists) virtually eliminated the fly from the entire cattle growing area by the simple expedient of releasing large numbers of male screw-worm flies that had been rendered sterile by radiation. Since the female screw-worm fly mates but once, a mating with a sterile male prevents the laying of fertile eggs. It does not necessarily follow that a similar treatment would eliminate all pests, but a study of the biology, including the mechanisms of heredity, of all species would contribute greatly to the solution of the problems that they present.

Opportunities for additional contributions of genetics are abundant. Further improvements in food crops and in breeds of domestic animals will help solve food problems. Better knowledge of hereditary diseases also is needed. Genetic studies of endless variety are in progress. To present an additional example: In another section of this chapter mention is made of the devastating effects of the chestnut blight fungus upon one of our most valuable forest trees. Efforts are in progress through genetic experiments to select blight-resistant chestnut trees, so that possibly once more one of the glories of the American forest may be seen.

Although many advances in genetics have been possible by experiments in crossbreeding of plants and animals, many opportunities are opening up in areas that owe their existence to the startling advances made by molecular biology. Information has accumulated in such volume and at such a rate that new agencies are being formed for the storage and subsequent retrieval of data in gene banks. The units of heredity, the genes, are made up of only four units or molecules—adenine, cytosine, guanine, and thymine, abbreviated as A,C,G,T, repeated over and over hundreds or thousands of times in varying sequence such as ACGTAGTCATGC.... The human genome, or the sum of all the genes in a single set of human chromosomes, is said to contain about three and a half *billion* such units. As the sequence of each new gene is determined, it can be compared with information already in the data bank.

Genome research received a major boost in the late 1980s, when Congress allocated significant funds to the National Institutes of Health, the Department of Energy, and the National Library of Medicine for the purpose of mapping the entire human genome. Known as the "Human Genome Project," the quest for the genome was expected to be an enormous effort, involving hundreds of scientists; numerous government, university, and private laboratories; and a number of computer and data centers. Scientists hope that the project can reach its goal by the early 2000s. That does not necessarily mean, however, that there will be a reduction of jobs for geneticists at that time. As molecular biologist Norton Zinder said when launching the genome quest, "When it's done, someone else will sit down and say, 'It's time to begin.'"

Some of the applications of genetics have aroused misgivings. Some people have expressed the fear, for example, that geneticists might inadvertently create new species of dangerous organisms. This problem has been discussed in conferences held by the scientists involved in such research, and they have agreed that they must act responsibly. In fact, there seems to be no indication that the worst fears expressed at the beginning of this research will be realized. Another source of unease is genetic testing of employees by employers. Some genetic traits might make certain employees predisposed to illness when they are exposed to certain chemicals, for example. Others might suffer genetic damage from exposure, which could be detected by monitoring, so genetic testing could possibly be used not only to screen prospective workers, but to detect any genetic changes after employment. Some have regarded this as an invasion of privacy, or as a possible means of discrimination if the practice is misused. However, the use of genetic testing is a growing practice, so employment opportunities for geneticists skilled in this area are increasing.

Industrial concerns employing geneticists also include pharmaceutical manufacturers; large producers of seed, especially hybrid seed corn; large producers of livestock and poultry; and large fur-breeding farms. Government laboratories that traditionally employ geneticists include the Department of Agriculture, the Fish and Wildlife Service, the National Institutes of Health, and many others. Agricultural colleges employ not only theoretical geneticists but also those concerned with practical applications in the management of livestock, poultry, or crops. They usually spend part of their time teaching and part carrying out research programs. Some universities have separate departments of genetics, but most of the employment opportunities in colleges and universities are in such departments as biology, botany, zoology, or microbiology.

From the above discussion, it is clear that genetics offers a variety of careers for those prepared to take advantage of the opportunities. A background in biology, chemistry, physics, mathematics, and computer technology would be helpful in giving the individual the widest choices.

BIOMEDICAL ENGINEERING

The biomedical engineer finds the answers to problems in patient care and clinical research using his or her training and experience in biology, medicine, and engineering. The results of the biomedical engineers work are usually devices or techniques that involve mechanical means of operation, but they can be as small as a computer chip that functions as an implantable drug pump or as large as a magnetic resonance imager, which when mounted on a tractor trailer truck can provide diagnostic services in the most remote locations.

Many exciting advances are being made in improvements in the care and rehabilitation of people who are handicapped because of the malfunction of some organ of the body that can be supported or even replaced by other organs or devices. Some of the more spectacular achievements in this field—such as a completely artificial heart—capture the public imagination and obtain much publicity. As interesting as such advances are, though, they actually represent only the tip of the iceberg of applications of new technology to human problems. Biomedical engineering is really a combination of medicine and engineering technology, and it is changing at such a rate that it is difficult to predict exactly what the career opportunities may be at the beginning of the twenty-first century.

Some applications of advances in physics and engineering to biological and medical problems have been in use for several years; these include implanted pacemakers for the heart, devices that generate electrical impulses to initiate heart beats when the heart's own pacemaker falters. Artificial kidneys in the form of dialysis machines are prolonging many lives, but future technology will greatly improve and simplify the use of such devices. Advances are being made in the replacement of limbs by prostheses that respond to the body's own neuromuscular impulses. Visualization of internal organs without the intervention of x-rays is also becoming possible through the use of ultrasound, nuclear magnetic resonance (NMR), and computed tomography (CT) technology. At a still more sophisticated stage are research efforts toward development of artificial sense organs, including even eyes.

The number of career opportunities available in biomedical engineering expands as research develops new applications of these technologies.

MICROBIOLOGY

Have you seen the latest on "the extremophiles?" No, this is not an episode of *Star Trek,* but it could very well be about extraterrestrials. Actually, extremophiles were the subject of a recent article in *The Scientific American* that described one of the many newly explored frontiers in microbiology…that of the microbes that thrive in conditions that we would call extreme.

The American Society for Microbiology was not exaggerating when it said, "microbiology is important from the core of the earth to the far reaches of outer space." Literally speaking, the word *microbiology* means the study of the very small—that is, organisms that cannot be seen without the aid of a microscope. Microorganisms include bacteria, so bacteriology is a branch of microbiology. Many people think of bacteria as "disease germs" since so many infectious diseases are caused by bacteria. A closer look, however, reveals the fact that of about fifteen hundred known species of bacteria only about one hundred cause disease. Many of the other fourteen hundred are of great importance to human beings, and we could not exist without their activities.

Microorganisms, for all their tiny size and simple appearance, produce many complex substances. Some of their products are harmful, but more of them are such useful substances as vitamins and antibiotics. As we have noted above in the discussions of biotechnology and genetics, the numbers of such useful products can be greatly increased by manipulations that include insertion of genes regulating specific chemical reactions within the bacterial cell. This is made possible by the fact that the structure of the DNA in a microorganism does not differ basically from that of the highest, largest, and most complex organisms. Some of the substances produced by microorganisms escape from the cells and produce profound changes in their environment. These include enzymes that convert complex materials into simple substances, helping in the disposal of waste and in making the materials available for reentry into the cycle of life. They fix nitrogen from the air; higher plants as well as bacteria use bacterially fixed nitrogen for building body proteins. In this way, much of the protein in our bodies comes directly or indirectly from bacterial action.

Microorganisms can endure an astonishingly wide range of conditions, and some are adapted to live in temperatures far below freezing, while others exist in the boiling water of hot springs. They can be studied from the point of view of taxonomy, morphology, biochemistry, physiology, ecology, genetics, or any other aspect of living things, just as are higher plants and animals. Algae, fungi, yeasts, viruses, rickettsiae, and protozoa are among the kinds of organisms studied by microbiologists. Many of these appear to be simple forms of plant life, but some cannot be readily classified as either plant or animal. Protozoa, meanwhile, belong to the animal kingdom. A special branch of bioscience, protozoology, deals with the protozoa.

Probably few people realize how widespread protozoa are; they are found everywhere there is moisture—in soils, oceans, lakes, and in the bodies of larger animals. Some of them possess hard skeletons and are so numerous in the oceans that the skeletal remains form thick beds of deposits—the White Cliffs of Dover, for example. Protozoa are important in the ecological balance; they consume bacteria, they help convert wastes into materials utilizable by higher organisms, and they are themselves consumed as an important link in the food cycle. A special group of students of protozoa are parasitologists. A few protozoa are harmful parasites, producing such diseases as malaria, amoebic dysentery, African sleeping sickness, certain venereal diseases, and others. Not all harmful parasites are protozoa, so many parasitologists devote their attention to larger parasitic creatures, such as roundworms. Thus, a parasitologist may be partly microbiologist and partly macrobiologist.

The viruses defy simple classification as plants or animals. They are too small to be seen with an ordinary microscope or to be held back by bacteriological filters, but virologists have devised many ingenious ways of studying them. Viruses have so many unique features that some virologists consider their specialty as a separate science and will hardly admit that virology is a branch of microbiology. Viruses are found in the cells of bacteria, plants and animals, and they may act as messengers, bringing in new genes. Some of the genes introduced into bacteria may alter the resistance to antibiotics. A single gene introduced into an animal cell may transform it into a cancer cell. Many plant diseases responsible for crop destruction are of viral origin.

Virologists are engaged in studying all these aspects of viruses, but some are trying to turn to people's advantage the predilection of viruses to invade animal cells and destroy them. Thus, viruses harmful to insects might replace toxic chemicals in sprays.

Mycology is a biological specialty that bridges the gap between microbiology and macrobiology. It is the study of fungi, which run the gamut in size from microscopic yeasts to huge puffball mushrooms. Mycology as a separate biological discipline will be discussed in the next section of this chapter.

Microbiologists contribute a great deal to the study of genetics, as we have seen in previous sections of this chapter. Some of these studies have produced new organisms, and the United States Supreme Court has ruled that such creations are patentable. Some strains of microorganisms appear to be especially sensitive to chemical compounds capable of inducing cancer. Such compounds cause mutations in the bacteria and this property makes it possible to detect potential carcinogens so that they can be removed from the environment.

Microbiology can boast some of the most illustrious names in the history of science and some of the greatest achievements for the benefit of human beings. Louis Pasteur, Robert Kock, Sir Alexander Fleming, and Jonas Salk are just a few of the great names that appear on the rolls of those who have contributed to microbiology. A third of all the Nobel prize awards in physiology and medicine have been bestowed upon microbiologists. Microbiology has so many practical applications that some microbiologists have complained that its value as a basic science has not been properly appreciated.

Microbiologists have been responsible for the development of vaccines, antiserums, and toxoids against a wide variety of diseases in human beings and animals. These include smallpox, typhoid fever, yellow fever, whooping cough, measles, influenza and polio. Some microbiologists hope that some day the common cold, cancer, AIDS, and others will be added to the list.

Many of our food and beverage products, including cheese, vinegar, wines and other alcoholic beverages, pickles, breads, olive and cabbage products, to name just a few, are dependent upon microorganisms, and microbiologists have contributed to the production and improvement of all of them. Microbiologists have also developed products for the preservation of meats, vegetables, and fish, and for the tenderizing of meats. Even oil geologists use microbiological information in their prospecting.

Diseases of animals are important targets of study by microbiologists. These include both the diseases that affect the animals and those that animals can pass along to human beings. According to the World Health Organization, about 150 animal diseases are important in human medicine; such diseases, called zoonoses, include some such as brucellosis and bovine tuberculosis, which have been conquered in most areas, but most of them remain to be controlled. In some areas of Africa, the cattle population could be doubled and the productivity of livestock markedly improved if just a single cattle disease—trypanosomiasis or nagana—could be controlled. Indeed, vast areas of the tropics have no domestic animals at all because of the presence of trypanosomes; and a few areas are virtually uninhabitable by human beings for the same reason. So there is plenty of work left for the microbiologists of the future!

The branches of microbiology concerned with disease-producing organisms include medical microbiology and veterinary microbiology.

Research on vaccines, antisera, and a variety of products of biological origin is performed by scientists in these areas. They are also making strides in investigating the role of viruses in cancer. A host of problems related to public health, such as the spread of infectious agents by insects, rats, or other animals; the sanitary control of water and food; and the disposal of sewage are the domain of the public health microbiologist. Agricultural microbiologists have a wide range of interests besides diseases of plants and animals, including such problems as the growth of microorganisms for use as food for human beings; yeasts and algae are among those suggested as possible future food sources. A hungry world also is awaiting the discovery of new organisms capable of extracting nitrogen from the atmosphere and turning it into a chemical form that can be used by plants. This can now be done by microorganisms in nodules on the roots of plants of the legume family; if a similar organism could fix atmospheric nitrogen for corn and wheat, for example, it would help solve the world's greatest food problem, namely, the need for high-quality protein.

Microbiologists teach and do research in nearly all colleges and universities and in many professional schools, including those of medicine, dentistry, public health, nursing, pharmacy, veterinary medicine, and agriculture. Private research foundations, government research laboratories and service agencies, public health facilities, hospitals, and agricultural experiment stations employ large numbers of microbiologists. Industrial microbiologists are employed by the food, chemical, and pharmaceutical industries in large numbers. These scientists are engaged in such areas as research, production, and quality control. The large-scale production of antibiotics and other complicated molecules can be done more effectively and economically by microorganisms than by the usual methods of synthetic chemistry in many instances.

Other industries that use the skills of microbiologists include wood products, paper, textiles, optical equipment, leather, and even electrical equipment. Their products are subject to microbial deterioration, and the microbiologist can forestall great economic losses by helping to control spoilage. In short, the man or woman trained in microbiology can have his or her choice of many careers. That training will involve basic courses in biology, chemistry, mathematics, and physics. Some colleges permit a major in microbiology; elective courses in microbiology and related subjects may vary depending upon the student's specific interest. If it is agricultural microbiology, the student will need botany and plant pathology, for example. If he or she is headed toward the chemical industries, a course in chemical engineering might be appropriate.

The top jobs in research and in university teaching will be held by those possessing doctoral degrees, as in most of the biosciences; but the variety of jobs available for microbiologists opens up many opportunities for those with baccalaureate degrees as well.

MYCOLOGY

Mycology is a word derived from the Greek word for fungus, and a mycologist is a student of fungi. Perhaps the most familiar examples of fungi are mushrooms, but the fungi also include molds, mildews, and yeasts. We are accustomed to think of all living organisms as belonging to either the plant or the animal kingdom, but the fungi have such unique properties that some mycologists like to insist that fungi belong to neither of these, but constitute a kingdom of their own. Certainly, they do not have the ability of most plants to convert carbon dioxide and water into substances that supply energy to animals eating the plants. For this process to occur, the green pigment, chlorophyll, is necessary. No fungus has chlorophyll, although some fungi may live in close association with algae containing chlorophyll so that the combination appears to be one organism—a lichen. Like most bacteria and animals, fungi are obliged to live upon organic matter previously made by green plants. If they depend upon dead organic matter, they are called saprophytes; if upon living organisms, they are parasites.

The requirement for previously formed organic matter to be food accounts in large measure for the immense practical importance that fungi have in the scheme of nature and in the economy of our civilization. Although a few of them cause disease in human beings, the majority are of benefit. They take part in the breakdown of dead trees and other plants, helping to return the dead tissues to the soil and the air, thus completing the cycle of growth and decay. Without them, trees would not rot when they fall, and even the autumn leaves would become an intolerable burden. Some of these activities bring them into conflict with people, as when fungi destroy the wood of houses, or when they attack food crops growing in the field, or when they attach themselves to our skin to cause athlete's foot or other, more serious diseases. The control of such undersirable activities presents a challenge and opportunity for mycologists.

The useful products of fungal activity are many, and with genetic engineering the list is growing. Medically useful products include penicillin, streptomycin, and other antibiotics as well as such steroids as cortisone and the hormones of birth control pills. Fungi turn milk into cheese and sugar into alcohol. They are abundant in the soil, on foods, in the air, on textiles and lumber—indeed, anywhere there is organic material and moisture. They are so widespread and their metabolic products are so useful that the discovery of new fungi is an important activity of some mycologists. The Mycological Society of America has estimated that there are over one hundred thousand different kinds of fungi known, and several hundred new ones are discovered each year. Many mycologists believe that there are more fungi and useful fungal products awaiting dicovery than have yet been found.

Fungi and their products are being studied by many specialists in other fields, including chemists, molecular biologists, physicians, ecologists, pathologists, and engineers. There are, then, many opportunities for collaboration between mycologists and others. Geneticists have learned much about heredity from studying the fungi, and cellular biologists and others have found much that is applicable to cancer research. In the applications of biology to forestry and agriculture, the mycologist collaborates closely with the plant pathologist, for the fungi are the chief cause of disease in higher plants.

The above description of mycology makes it apparent that there are many kinds of opportunities open for persons skilled in this field. The importance of mycology is becoming increasingly recognized, and many colleges and universities offer courses in mycology that did not do so a few years ago. These schools, of course, employ mycologists as teachers and research workers. Such professional schools as medicine, forestry, and agriculture also utilize the skills of mycologists. Medical research laboratories, whether in private research institutes, in the pharmaceutical industry, or in installations of the public health service, employ mycologists. In such institutions, fungi are studied not only as disease-producing agents, but many forms are cultivated and studied for their value in producing antibiotics and other medically useful products in ways that chemists cannot economically duplicate in the test tube. Industries and laboratories interested in food production, leather, textiles, and forestry products also have need of mycologists. Chemical manufacturers also seek to discover means of preventing spoilage of manufactured goods by devising and testing chemical substances that will control or kill fungal growth; mycologists are employed for testing such substances.

It should be pointed out that many of the positions mentioned above also may require knowledge of some field other than mycology and often only a narrow portion of mycological information is used in some particular job. For example, master bakers, vintners, and brewers may know a great deal about yeasts, but not neccessarily enough about other fungi to qualify as scientific mycologists. Other examples of the application of the narrow spectrum of mycological information include pharmaceutical chemists who know all about fungi that produce antibiotics and other useful medicinal compounds. Relatively few of such specialists have a broad knowledge of mycology. Most *bona fide* mycologists belong to the Mycological Society, and a survey of the membership revealed the fact that about 87 percent of the members are employed either in teaching and research positions in colleges and universities or in laboratories of state and federal governments. The society points out that the mycologists with the broadest opportunities in industry also have some skills in other professions. For example, a mycologist with some knowledge of industrial chemistry would have doors open for more places than would someone trained in either field alone.

A mycologist should have first of all a base in biology in general, which in turn involves the basic sciences of mathematics, chemistry, including organic chemistry and biochemistry, and physics. Most mycologists begin to develop their interest in fungi when they are about halfway through college, and the fully professional mycologist will obtain advanced degrees in graduate school. As pointed out above, however, there are many jobs available in specialized categories that do not require that one be a complete mycologist. If one does have both the special and the complete training, however, many more opportunities will be available.

SYSTEMATIC BIOLOGY

Systematic biology—also referred to as systematics or biosystematics —is a field that may have special appeal to those who enjoy orderliness and who like to be independent. Systematics may be defined as the study of the different kinds of organisms that exist, as well as the kinds that existed in the past and are now extinct. It includes a description and classification of organisms and of the relationship between them, as well as the changes that have occurred in them during past generations. One aspect of systematics, taxonomy, studies the appearance and structure of a plant or animal, determines how it differs from others, describes its surroundings, classifies it in its proper place in the plant or animal kingdom, and names it. The aspect of systematics concerned with how organisms change over the course of generations and what influences are at work on them is called evolution. It seeks to determine where a given species came from, what processes are acting upon it to keep it as it is or to change it into something else, and whether we can predict what forms its future may take.

Systematists are not merely concerned with appearances; they use the tools supplied by other disciplines. For example, subtle differences in the structure of DNA from one species to another may give clues of their genetic relationships. The systematist may make original observations or use data supplied by other biologists, such as physiologists—the microscopic cell structure, the biochemical reactions taking place within living tissue, the breeding behavior, the geographical relationships, the interplay between the organism and its environment, or any other data that may help to establish the position of the organism within the plant or animal kingdom.

An ongoing project at the National Museum of Natural History is the study of the genetic diversity of endangered species as well as the genetic material from museum collections and fossils. These studies will help resolve questions about nature, history, and evolution.

Systematics is really a collection of many specialties, although there are a few general systematists. For example, a mycologist studies yeasts, molds, mushrooms, or other fungi. A dendrologist specializes in trees, and an algologist concentrates on the small and simple green plants known as algae. An entomologist devotes his or her attention to insects, an ornithologist to birds, a herpetologist to reptiles and amphibians, an ichthyologist to the fishes, and a mammologist to mammals. It would be the exceptional student who deliberately chose to pursue one of these specialties before beginning the study of biology, as one's interest in a restricted field develops during the course of study. Indeed, there are few jobs specifically labelled herpetologist or ichthyologist. Individuals with such specialized interests are biologists first of all, and they follow their specialty while teaching the usual biology courses. The American Society of Ichthyology once put it this way: "Nobody is ever hired merely to teach ichthyology…the young ichthyologist may…find that almost all of the available openings are for teachers of introductory courses like general biology." However, there are some exceptions to this. A limited number of openings for curators of ichthyology, herpetology, or other specialties do occur from time to time—mainly in such places as museums, zoos, aquariums, and fisheries laboratories.

The glamour and importance of certain fields of experimental biology, including biotechnology and some of the biomedical sciences, may have overshadowed systematic biology in the eyes of many, but it is well that we recall the enormous contributions made by systematists to the quality of life that we enjoy. There is an urgent, and indeed critical, need for a thorough understanding of biological interrelationships if our civilization is to survive. In this respect, systematics interacts with and overlaps other biological disciplines, especially ecology and genetics. A few examples may be instructive.

During the present century, human life expectancy in many parts of the world has increased more than it had in all the thousands of years since the dawn of history. Our lives are not only longer but healthier. We are aware of the importance of medical research in this advance, but the contribution of systematic biologists to health and longevity is less well recognized. Taxonomists have studied and described harmful bacteria, plants, and animals; they have determined their relationships to other organisms, including human beings; and they have described the position that each species occupies in the natural scheme of things (that is, their ecological niche). Without the foundation of basic knowledge provided by systematic biologists, the great advances in health science and medicine for which we all are so grateful would have been immensely more difficult if not impossible. This is true not only in the study of organisms likely to cause disease, but also in promoting the discovery of antibiotics and of improved sources of essential nutrients including vitamins and proteins. In the 1920s, two systematic botanists spent two years in Asia. They spent $50,000 of American taxpayers' money but brought back soybean types that now produce crops worth $2½ billion annually to American farmers. Many plants and animals transported from one country to another prove to be valuable additions to the economy of the country into which they are introduced. Some of them, however, prove to be pests. An example of the latter is the prickly pear cactus in Australia. After it had ruined some fifty million acres of valuable grazing land, the help of the systematic biologist was sought in finding means of controlling the pest. An intensive study of all the information available on the prickly pear and on the plants and animals living in association with it revealed 160 different kinds of insects that were injurious to the cacti (50 of them were new to science). When some of these were introduced into Australia, one of them, a small moth from Uruguay, proved to be so damaging to the cacti that within seven years practically all the area formerly occupied by the prickly pear was reclaimed for grazing. This triumph for the systematists cost less than $1 million, as compared to the estimated cost of eradication by spraying or by mechanical means of more than $500 million.

The outstanding success of American agriculture is based upon the introduction of plants and animals native to other countries. A program of research on the systematics of species of potential economic importance is necessary if further useful introductions are to be made. Only by learning as much as possible about the biology of foreign species can a scientific prediction be made as to whether a new plant might be useful like the soybean or might be a troublesome weed like crabgrass; whether a foreign fish might be desirable like the brown trout or undesirable like the German carp; whether an introduced bird will make an honored place for itself like the ring-necked pheasant or will become a nuisance like the starling.

There are many examples of introduced species that have become pests. Some of these introductions were made purposely but without adequate study of the relationships between the native and the exotic species. Others were made through accident or carelessness, often including no-longer-wanted pets. The numbers of such releases are astonishing; *annual* imports amount to more than one hundred million fishes, more than half a million birds, and more than one hundred thousand mammals! There is an urgent need for better systematic knowledge of many groups involved in these imports.

Systematists have given valuable assistance to programs designed to encourage biological control of pests and diseases. They aid in ecological studies in many ways, such as identification of organisms in water to judge the degree of water pollution. Physicians and toxicologists in hospitals and public health laboratories employ systematists as consultants to aid in the identification of such toxic plants and animals as poisonous mushrooms and snakes. They participate in the discovery of plant strains resistant to disease. They identify fossil species of plants and animals in sedimentary rocks as an essential part of oil and mineral prospecting. (Does it surprise you that oil companies employ systematic biologists?)

From all of these examples, you can see that there is scarcely any field of biology that offers such a wide variety of career opportunities. Systematics involves both field and laboratory work and deals with all manner of terrestrial and aquatic habitats. Systematic biology offers opportunity to travel both in the civilized and the wild parts of the earth. Systematists also can relate their professional interests to many other technical and academic fields, including such nonscience subjects as history, logic, and classical languages.

Systematists find employment as teachers in secondary schools, colleges, universities, and agricultural colleges. Museums, botanical gardens, arboretums, zoos, aquariums, private research foundations, and various local, state, and federal agencies employ systematic biologists. The federal government has been employing them since the time of the Lewis and Clark Expedition of 1803–04, and most of them are in the departments of Agriculture or Interior. Students in general have been so impressed with the supposed glamour of research in the health sciences that relatively few have considered systematics as a potentially interesting and useful discipline. Now, when a number of important national and international research programs relative to our environment are under way, it is important that there should be no shortage of systematic biologists.

ENTOMOLOGY

During past ages, various forms of life predominated at different times. For example, during the Jurassic period some 150 million years ago, reptiles dominated the land, and we speak of the Age of Reptiles or of the Dinosaur Age. In like manner, we may consider our own time to be the Age of Mammals, but in another sense it could just as well be termed the Age of Insects. More than three-fourths of all the species of animals in the world today are insects or their close relatives. From the point of view of human beings, many are pests spreading disease or competing for food or destroying desirable plants, while others are useful in pollinating trees, crops, and other plants or providing food for us or helping to destroy pest species. Whichever they do, insects affect the lives of every person. Entomologists are the biologists who study them.

Entomology overlaps with many other biological disciplines. Although the description and classification of insect forms have been going on for more than a century and hundreds of thousands of insects have been described and named, scientists estimate that there may be as many as ten million types of insects that have not yet been discovered. The identification and classification of insects are fundamental to all other research in entomology.

Those biologists who study the intimate nature of the physiological and biochemical reactions taking place within cells or even at the molecular level find that insects provide an abundance of useful and interesting material. The insect physiologist makes discoveries that are useful in many areas—sometimes even in human medicine. An important part of research in entomology is the study of insect control; that is, research aimed at reducing or eliminating the harm done by insect pests. Contributions to these problems are made by those in both basic and applied research. The economic entomologist is especially concerned with reducing the damage done by insects to crops, forests, food in storage, and other interests of people.

In the past, most attention has been paid to insect control by means of poisonous chemicals, so the work of the entomologist has overlapped with that of the toxicologist. Now, many are interested in the use of chemical substances that are not systemic poisons: they might produce abnormal growth or behavior of the insects, they might attract insects to traps, or they might render insects sterile so they cannot reproduce. These goals are helped by research on the physiology, reproduction, and behavior of insects, and the knowledge gained in this research is often of interest in other fields of animal and human biology.

Another avenue of approach to insect control is known as biological control. This involves the study of other insects or other forms of life that prey upon or destroy the insects we regard as pests, as well as the study of insect diseases—bacteria, viruses, or fungi that attract insects. The work of the entomologist thus may be closely linked with that of the microbiologist, the virologist, or the mycologist. No form of life can be destroyed without affecting what is sometimes called the balance of nature. Hence, the entomologist engaged in insect control will have occasion to collaborate with the ecologist and the wildlife biologist.

It would be difficult to mention a field of biology offering a greater variety of employment opportunities than entomology. Most teachers of entomology are hired by colleges and universities—especially colleges of agriculture and veterinary medicine. Industries that employ entomologists include producers and processors of food, the chemical industry interested in chemicals for insect control, and the lumber and pulp industries, which utilize the skills of forest entomologists. We see constant reminders to help prevent forest fires, but few realize that insects destroy more timber each year than do forest fires.

The production and storage of food and fiber is of the utmost urgency for the expanding world population—now increasing by about 1.6 percent annually, which would mean that the world population will double by the year 2028. Meeting this challenge would be greatly simplified if insect damage were eradicated, for it has been estimated that fully one-third of all production is destroyed by insects. Success or failure in controlling insects could well make the difference between starvation or survival in much of the world, especially the developing countries. Along with this, insect-borne diseases make life miserable for millions, and indeed the areas at present that are uninhabitable or nearly so would be accessible if the insects could be controlled.

There are, then, tremendous challenges in entomological research, but the application of research findings also offers many employment opportunities and is of immense importance. For example, in agriculture, knowledge gained by the research scientist is transmitted by extension workers to farmers. The research findings are also passed on to the next generation by teachers. The private sector also employs many people with skills in entomology. Major chemical companies employ entomologists skilled in toxicology and insect control. Although in the past mistakes have been made in the application of pesticides, there is growing realization that skilled personnel are greatly needed in devising and applying a balanced system of pest management. In addition to agricultural pests, there is a large industry in the United States engaged in control of termites and other pests that attack dwellings and other structures.

The demand for skills in the control of pests is great. Employment as a technician in pest control does not require a college degree, but some specialized education is needed for advancement and both two-year associate degrees and four-year baccalaureate degrees in pest control may be obtained. Some of this training is available by correspondence.

Many entomologists are employed by state and federal governments in research laboratories and in biological survey work. Others are extension entomologists, helping farmers, growers, nursery workers, and householders wherever insects present problems in everyday living and commerce. Health agencies employ medical entomologists—that is, those specializing in insects responsible for the spread of disease. Agriculture experiment stations, plant inspection agencies, mosquito control boards, conservation agencies, and museums are a few of the other sources of employment opportunities. Several hundred inspectors are employed by the Animal and Plant Health Inspection Service of the U.S. Department of Agriculture to identify pests that hitchhike to the United States on imported animals and in tourists' pockets. These entomological inspectors perform an important and useful service. For example, inspectors have intercepted bark beetles that help spread Dutch elm disease and woodborers in crating and packing material, as well as fruit hidden in flight luggage containing larvae of fruit flies capable of threatening an entire industry such as citrus growing.

The largest organization of professional entomologists is the Entomological Society of America, with about ninety-two hundred members. The society encourages the highest standards among its members, promotes communication of information, and maintains a permanent office that answers inquiries about educational requirements and job opportunities.

MARINE (OCEANOGRAPHY) AND AQUATIC BIOLOGY

Some biologists especially concern themselves with plants or animals that live in a watery environment. They may find fascination in a study of the classification, mode of life, functions, or adaptations of organisms inhabiting the sea—marine biology—or fresh water—aquatic biology. Few biologists will have interests broad enough to cover all these aspects thoroughly; one may be especially interested in the physiology of fishes, another in the systematics of marine algae, yet another in the structure or embryology of a particular group—for example, the minute invertebrate animals that form a part of the plankton (the small floating or swimming forms that are the basis of the food chain for all the larger creatures.

The economic importance of marine biology is being increasingly recognized. Most of us have read about the threat to the survival of human beings that has been posed by the often-mentioned population explosion. For many years, numerous observers have foreseen serious food shortages in widespread areas of the earth, and some have forecast worldwide famine. Recent reports from several regions have borne out some of these unhappy predictions. There is no question that population is increasing faster than agricultural production in the world as a whole. Marine biology may offer one bright spot in this gloomy picture. Food from aquatic and marine organisms contributes protein, vitamins, and minerals to human nutrition—the very essentials that people in developing countries need the most. The need can be satisfied only by the results of continuing research efforts by marine and aquatic biologists.

The increase in worldwide consumption of fish is not only for human food. Commercial livestock feeds and fertilizers demand an increasing share. The improved efficiency of fishing techniques has put great strains upon the ability of the sea to keep up. The American Fisheries Society has stated that demands for fish products have threatened with irreversible depletion stocks of fish previously thought to be inexhaustible. Nor is this the only area presenting challenges to the fisheries biologist. The estuaries of large rivers and coastal marshes and wetlands form essential links in the chain. They are breeding areas for many species of fish and of the smaller organisms that supply the fish with food.

Recreational fishing also is an important factor. Such is its growth that it is estimated that by the year 2000 some sixty-three million anglers will spend 1.3 billion days fishing! Both commercial and recreational activities are threatened by overexploitation and by deterioration in water quality.

Relying upon fishing to produce needed food is equivalent to the method of prehistoric people. The aquatic equivalent to agriculture is aquaculture, and only through this procedure can we realize the full potentialities of aquatic organisms. Actually, aquaculture is not new; it is widely practiced in some countries. Indeed, a treatise on fish culture was written in China as long ago as 475 B.C., and four-fifths of the world's aquaculture is still practiced in East Asia. Among the species of aquatic and marine animals widely cultivated are salmon, trout, edible carp, catfish, oysters, and shrimp. Aquaculture has begun to assume an important role in the United States as many large firms have entered the field. However, the fledgling industry still produces only about 300 million pounds of food annually, which accounts for about 8 per cent of America's seafood consumption.

Energy conversion by water-dwelling animals is far more efficient than by land animals; that is, for each pound of food energy they consume, a greater portion of it is converted to meat suitable for human food. The explanation for this remarkable circumstance is twofold: temperatures in the water environment do not fluctuate wildly as they do in the air; hence, aquatic organisms do not have to waste much energy in combating cold or in cooling their bodies during hot spells. In addition, the water they live in weighs nearly the same as does an equal volume of their own bodies; hence, they do not need to spend large amounts of energy for antigravity purposes.

For these reasons, the yields achieved in water far outstrip those on land. Fertilized catfish ponds in the southern part of our country may yield as much as 4,500 pounds of fish per acre per year. This figure is dwarfed by the crops harvested from salt water. Some figures quoted by workers in ocean research laboratories range from 51,000 pounds of meat per acre annually in Japanese oyster beds to the fantastic figure of 268,000 pounds per acre per year in Spanish mussel colonies. To give some idea to the meaning of these figures: a good yield of small grain on a farm may be on the order of 100 bushels per acre. At 60 pounds per bushel, this would be about 6,000 pounds of grain per acre. When this is fed to animals for producing meat, it will yield about 600 pounds of unprocessed meat.

The full potentialities of aquaculture cannot be realized without the contributions to knowledge that only the biologist can supply. For example, the biosystematist could study the relationships of the species concerned and perhaps find useful species that are not yet used for food. The systematist also would be in a position to study the many lowly members of the food chain in order to provide most effectively the food for the species being cultivated. The systematist also could supply information to aid the geneticist in selecting species for hybridization, and the geneticist would supervise selective breeding and hybridization. A program at the University of Washington in Seattle using rainbow trout as subject produced a breed that grows to a weight of six to seven pounds in the time it takes an ordinary trout to grow to six or seven ounces. This new breed also tolerates water temperatures that the wild trout cannot endure.

Similar studies focus on the valuable food fish, the salmon. The studies include the problems inherent in migratory and breeding habits of the anadromous fish; that is, those fish that live in the sea but spawn in fresh water. Considerable progress has already been made in the adaptation of aquaculture techniques to selected strains of such fish, but aquatic biologists are vigorously pursuing improvements. Subtle changes in temperature, salinity, and other properties of an aquatic environment can make great changes in its suitability for a given species.

There are many unsolved problems in aquaculture. For example, although oysters and shrimp are widely cultivated, experiments are needed on the genetics of these creatures. Techniques for growing lobster and abalone have recently been developed, but some companies have found attempts at cultivation of several species of seafood animals to be unprofitable, and some to have given up attempts. Others, however, are very successful. Thus, the need for additional research is apparent. Studies of the physiology of marine invertebrates need to be done. Such studies would include nutrition and food supply for the animals under study, reproductive and endocrine physiology, the diseases that might threaten the venture, and all other aspects of their lives.

From this discussion, it is clear that there is plenty of work to be done by anybody choosing marine or aquatic biology for a career, and the potentialties for aid to humanity are great. Aquaculture, however, is only a special application of marine biology and is by no means the only source of career opportunities in these fields. Departments of oceanography in universities conduct studies in marine biology as well as in the chemistry, physics, and geology of the marine environment. The biologists in such departments are concerned with all aspects of the systematics, biochemistry, and physiology of plants and animals in water. Many departments of botany and zoology also employ biologists interested in aquatic and marine organisms. Institutions and agencies interested in the impact of society's pollution tendencies upon watery habitats also offer employment opportunities. These effects are often difficult to study because of the exceedingly complex effects of currents in bays, fjords, estuaries, as well as in the open sea and in streams. The biologist with a good foundation in mathematics and in computer technology is in a particularly good position to tackle these difficult problems.

Marine and aquatic biology offers a seemingly unending variety of occupations, and with advances in technology the list of possible careers expands. There are opportunities for outdoor occupations and a wide variety of indoor employment, or a mix of the two. To mention just a few of the openings that have been filled within recent years:

- Monitoring salmon migration for a state agency
- Inspecting fish handling aboard a processing vessel

- Serving as agent for the marine advisory service of NOAA (National Oceanic and Atmospheric Administration)
- Analyzing water samples for culture of shellfish
- Managing fish hatcheries
- Surveying fish population
- Supervising a fish farm
- Investigating effects of volcanic eruptions on lakes and rivers
- Preparing environmental impact statements
- Managing a stream for a private club

An especially active field with many job openings deals with the handling and processing of fish and shellfish for human consumption. These openings require a knowledge of food science as well as marine biology. Food technologists in this field may be involved not only with the familiar food fish such as salmon and cod, but also with species available for food, but not much utilized—such forms as shark, squid, alewives, and eels.

Agencies of the U.S. government that employ those trained in fisheries science include: Department of the Interior—U.S. Fish and Wildlife Service, National Park Service, Bureau of Reclamation; Department of Commerce—National Oceanic and Atmospheric Administration; Department of Agriculture—Farmers Home Administration, Forest Service, Soil Conservation Service; State Department—Bureau of Oceans and International Environmental and Scientific Affairs.

Some international agencies, such as the Food and Agriculture Organization of the United Nations, also employ marine biologists. Special laboratories devoted to research in marine and aquatic biology are located on the Atlantic, Pacific, and Gulf Coasts, and some are inland. A few are independent, such as the well-known Marine Biological Laboratory at Woods Hole, Massachusetts. Others are connected with universities, such as the Rosenstiel School of Marine and Atmospheric Sciences at the University of Miami and the Scripps Institute of Oceanography of the University of California at San Diego. Some university laboratories are located in close proximity with other agencies; for example, the College of Ocean and Fishery Sciences of the University of Washington in Seattle includes a School of Fisheries that offers undergraduate degrees in fisheries science or in food science. Nearby is a laboratory operated by the State Fisheries Department as well as the Northwest and Alaska Fisheries Center of the National Oceanic and Atmospheric Administration of the U.S. Department of Commerce.

These institutions and others like them offer careers to biologists primarily interested in research. In those supported by universities there are ample opportunities for teaching as well as for those who enjoy routine technical work, such as chemical analysis or plankton sampling. Some of these institutions support oceangoing research vessels—floating laboratories that range all over the world and offer especially interesting opportunities for a limited number of biologists. Areas of research in marine biology make a long list; they may embrace ecology, water quality studies, culture of organisms for food for fishes as well as culture of fish, physiology, pathology, limnology, computer modeling, and many other specialties.

The person considering marine or aquatic biology has a wide choice of colleges and universities where education and training can be obtained. Some schools offer undergraduate programs in fisheries science, but the American Fisheries Society emphasizes that before specializing, students should first obtain a thorough background in biology, including anatomy, microbiology, genetics, and biochemistry. Mathematics, computer science, and statistics also should be included, and one must not forget communication skills, English, and the humanities. Sometimes there are opportunities for summer jobs, which not only provide valuable educational experience, but also will help the student decide whether this is a career he or she wishes to pursue. There are more than a hundred colleges and universities offering courses related to fisheries sciences at the undergraduate level. They include such subjects as fisheries biology, fisheries management, freshwater biology, fish physiology, fish genetics, fish behavior, fish diseases, water quality management, and many others. The American Fisheries Society maintains a list of educational institutions offering such courses. Additional sources of information on colleges offering degrees in marine biology and related fields include *The College Blue Book* and the *Index of Majors*.

Although the most challenging career opportunities in marine biology research are available primarily to those with advanced degrees, there are large numbers of positions open for those interested in practical applications. Several commentators have urged that the exploration of the inner space of the earth's oceans be given a priority at least as great as the glamorous exploration of outer space. As the importance of knowledge of the seas becomes more widely recognized, career opportunities in marine science, both biological and physical, will continue to increase.

SOME SPECIALTIES IN PLANT SCIENCE

There are many varieties of botanists. They are the biologists who are especially interested in plants. Plant systematists or taxonomists describe, classify, and study the evolutionary interrelationships of the nearly endless variety of members of the plant kingdom. Plant morphologists are fascinated by the form and structure of plants; plant ecologists with their environmental relationships. There are no hard and fast lines between these various specialties. It is clear that the systematist, for example, must have an understanding of morphology and ecology. Plant physiologists are primarily concerned with the normal functioning and behavior of plants, how the plant absorbs energy and what it does with it, for example. The highly important problems of plant disease engage the attention of the phytopathologist (or plant pathologist). Some botanists specialize in certain groups of plants; for example, the bryologist studies mosses and liverworts, and the dendrologist concentrates attention on trees.

Other plant scientists use their training in practical applications of their skills; they may become foresters, horticulturists, or agricultural scientists. Careers in botany, then, may be quite varied. A few of them will be considered in the following sections of this chapter.

PLANT PHYSIOLOGY

One could scarcely imagine a science of greater importance to humanity than plant physiology. This becomes apparent when one reflects upon the fact that all life upon earth, human beings included (with the exception of a very few unusual forms of bacteria), relies upon the sun as its ultimate source of energy, and that all animal life—including people—is utterly dependent upon the physiological activities of plants to convert the sun's energy into a form that suits our own needs. This is done, of course, by photosynthesis, the process by which the pigments of plants, especially chlorophyll, absorb the energy of the sun's rays and use this energy to convert carbon dioxide from the air and water from the soil into sugar. The sugar then may be converted into starch or fat or other substances, and with the addition of nitrogen from the soil into protein. The products of these reactions are the basic ingredients for all the metabolic processes of our own bodies—the processes that enable us to exist.

The increasing population of the earth will make continually greater demands upon our ability to control the growth and development of plants for improved food production; the plant physiologist is one of the key persons in this struggle. We need, for example, a more thorough understanding of the mechanism of photosynthesis, for the chemist still cannot duplicate photosynthesis in the test tube. A large percentage of the efforts of plant physiologists is devoted to the study of plant metabolism, including photosynthesis, as well as the mechanisms by which the plant absorbs and transports water and minerals. A knowledge of the utilization of water by plants would be valuable in those parts of the world where water is a precious commodity in short supply. The increasing uses of water for industry in our own country and the resulting effects upon the reservoir of underground water stresses the importance of learning the role of plants in the overall cycle of water. More thorough knowlege of mineral metabolism in plants could help in learning how various plants adapt to soil of differing mineral content. And, of course, better understanding of photosynthesis might help immeasurably in increasing the manufacture of basic food substances; at the present time, only plants are able to make them.

Discoveries of plant physiology that have already been of value include the chemical substances—plant hormones—that regulate growth, cell division, and the differentiation of the various organs of the plant, such as roots, leaves, and flowers. Such knowledge has been useful in regulating flowering and the setting of fruit and in the destruction of weeds by the stimulation of abnormal overgrowth. The understanding of how these substances act is incomplete, and more needs to be known about them.

One expanding facet of plant physiology is that of environmental physiology. In the natural course of events, plant growth may be affected by salt in the air; saltwater spray may affect the kinds of plants that thrive near the seashore. The effect is seen for a mile or more inland and even further during hurricane conditions. A comparable action is produced by salt placed on highways for ice control. The increased air pollution of recent years has exposed plants to gases and particulate matters for which they were not adapted, and there have been many instances of injuries to plants from these sources, both economic and aesthetic. The sources include the burning of fossil fuels, incomplete combustion of coal or wood, and a great variety of industrial emissions. Injuries appear whenever a certain concentration is exceeded; different species of plants respond to different concentrations and exhibit different signs and symptoms. The identification of the trouble is a highly specialized field. In these studies, the plant physiologist, the ecologist, and the meteorologist cooperate with one another.

The solution to these unsolved problems relating to plant physiology will be sought by those plant physiologists who are engaged in basic research, but there are also many involved in applying basic concepts to practical applications. About two-thirds of all plant physiologists work in universities, most dividing their time between teaching and research. The remainder are employed in government laboratories, in industry, or by private research foundations. The American Society of Plant Physiologists has estimated that research in plant physiology is about equally divided between basic and applied. Plant physiologists interested in applied research find many ways in which their chosen science is of practical benefit. If they are in industry, for example, they may help to develop new chemicals or new ways of using chemicals for such things as controlling plant growth, increasing plant production, or destroying weeds.

The plant physiologist may deal with organic chemistry, physics, genetics, plant pathology, molecular biology, and other disciplines, but the job is to analyze and coordinate this knowledge. The plant physiologist may deal with a small part of the plant, but the ultimate goal is the understanding of the entire organism. It is no wonder that plant physiologists are found in all sections of the country, in every major university, and in all the agricultural research centers.

The education and training of plant physiologists do not differ markedly from those of other biologists during the undergraduate years. Although a basic understanding of biology is necessary, some plant physiologists have expressed the view that the physical sciences—mathematics, chemistry, and physics—are so important that physical science majors make good candidates for concentration on biological studies in graduate school leading to an advanced degree in plant physiology. There are, of course, many jobs in such places as plants' physiology laboratories and greenhouses that do not require advanced degrees, but the top jobs usually are held by someone with a doctorate.

It may be seen that plant physiology forms a connecting link between basic biological research on the one hand and agriculture, forestry, and similar applied disciplines on the other. Biologists trained in this field will find a variety of outlets for their talents.

PLANT PATHOLOGY

Closely allied to plant physiology is plant pathology (or phytopathology), which shares with its sister science the quality of being of supreme importance for the maintenance of our civilization. While the plant physiologist contributes discoveries that enable us to understand plant growth, development, and behavior with corresponding improvements in food production, the plant pathologist provides knowledge that enables us to control the plant diseases that continually threaten to wipe out all the gains made by agriculture, horticulture, forestry, and similar disciplines. The success or failure of the plant pathologist determines not only how well we eat, but whether we can survive.

Without the work of the plant pathologist there would be few major crops grown in the United States, and our great fruit and vegetable industries would collapse. The threat of plant disease is not an idle one. Only a generation or so has passed since rust threatened to destroy the wheat-growing industry of the United States, and wheat rust still causes millions of dollars in losses each year. The entire corn industry in the United States was threatened by blight as recently as 1970. A potato blight in the 1840s caused death by starvation of a quarter of a million people in Ireland and was largely responsible for the mass migration of Irish people to America. One of the most beautiful and useful trees of the original American forest was the chestnut. Its fruit was a delicious nut; its wood was valuable for furniture and cabinetry; its growth beautified the landscape and even inspired poets. Within a couple of decades after the introduction of the chestnut blight fungus into this country, the American chestnut had almost completely disappeared. Many millions of our beautiful elms have been killed by the Dutch elm disease.

These few examples serve to illustrate how plant diseases continually threaten to cause critical shortages of food, fiber, and building material. The near miracles wrought by plant pathologists in the past have not abolished the threats, and there remain challenges as great as any that have previously been met. Plant diseases still cause billions of dollars in losses every year—fruits, grains, vegetables, flowers are attacked by bacteria, fungi, viruses, and other pathogens, as are animals. The tools that will be employed to meet these challenges will include the breeding of plants resistant to disease and the use of chemicals, including antibiotics and plant hormones, for combating disease. The plant pathologist thus cooperates with the plant physiologist, the biochemist, and others.

The American Phytopathological Society has estimated that there are more than fifty thousand destructive plant diseases. This science is growing in importance and plant pathologists find employment in many diverse areas. Courses in plant pathology are offered in universities in every state in the Union and in Puerto Rico. Teachers trained in plant pathology are employed in teaching such courses not only in departments of plant pathology but also in other biology departments, as well as in agricultural schools, in extension courses, and sometimes in secondary schools.

Research positions are available in state agricultural experiment stations as well as in the U.S. Department of Agriculture and in some other government departments, including the foreign service.

All major industrial firms engaged in the production of agricultural chemicals employ plant pathologists, and the federal government's program of technical aid to foreign nations engages many plant pathologists. Other phytopathologists work in conservation projects. These may be in wide-open spaces, ranges, national parks, and forests, or in cities working with grasses, trees, flowers, and ornamental plants. Some are becoming freelancers, that is they go into private practice as consultants on plant disease or start up their own services for testing or disease control. They are doctors of plant disease, just as veterinarians are doctors of animal disease, or physicians are doctors of human disease. Their numbers are small, but growing.

There are positions in plant pathology for those with diverse levels of education. A baccalaureate degree in plant pathology is offered by some schools whose graduates find jobs with food processors, manufacturers of agricultural chemicals, and similar employers. They also may become federal or state plant inspectors, or they may teach in secondary schools or in two-year colleges. Those who continue their education through the doctorate will be equipped to handle the most challenging problems in the field and to qualify for the highest positions. The person who chooses plant pathology will be in a position to help humanity and all the animal kingdom in the process of helping the members of the plant kingdom.

ECOLOGY AND WILDLIFE BIOLOGY

Until recently, ecology was a strange word to many people, but the rise in environmental consciousness has opened many doors for biologists interested in conservation and in protecting us from environmental damage. The study of the interaction between living things and their environment is not new but has recently assumed an importance in the popular mind not previously accorded to it. With the vastly increasing impact the people have had on the environment, it is important for our survival that we understand what we are doing to the world in which we live. It is encouraging that within recent years many people are becoming aware of the concepts of ecology, bearing out the prediction made in the late 1960s by Prince Bernhard of the Netherlands, who was president of the World Wildlife Fund when he predicted that ecology would become a household word to those who were then ignorant of its meaning.

It is difficult to understand what latecomers human beings are upon the earth. Several years ago, Paul B. Sears, writing in *American Scientist,* pointed out that if the entire history of the earth were written in a book with each page recording a million years, we would have an immense volume of four thousand pages. The beginnings of life would be recorded in the middle of the book, but human beings would not appear until the last page (recent discoveries indicate that this might be pushed back a page or two). The beginnings of agriculture would not be found until very nearly the end of the last line on the last page. If we think of what that means, we can see that the laws of nature became fixed long before human beings appeared on the scene. And there is nothing to suggest that people are exempt from the operation of those laws. The future of our species depends upon maintaining a balanced harmony with nature.

The extinction of the dinosaurs was a very impressive event in geological history, but if we reflect that a thousand species of dinosaurs became extinct over a period of a million years, we can see that on the average only one form disappeared in each one thousand years. By contrast, activities by human beings have destroyed more than two hundred species of birds and mammals within the past two or three centuries. Many others have been so drastically reduced in numbers that they may be doomed in spite of efforts to save them. It is not necessary to destroy the last member of a species in order to eliminate it, for there is a certain critical level below which a species fails to reproduce.

History is replete with examples, both ancient and modern, of adverse effects upon human culture when the principles of ecology are flaunted. The environment was able to absorb everything that we could do to it so long as technology was in a relatively primitive state and our numbers were not too great. But now it is threatened with a technology that includes even the potentiality of destroying all life and with an increase in the rate of population growth that, if not curbed, will render ineffective all other efforts at preservation of an environment in which most of us would like to live. As Dr. Vinzenz Ziswiler, of the Zoological Museum of Zurich, said, "The protection of nature must have the preservation of mankind as its ultimate end...When man continues to destroy nature, he saws off the very branch on which he sits."

Some of the problems addressed by the ecologist, then, are of basic importance, and many international programs are heavily oriented toward ecological problems. Ecologists often need some fields other than their specialty; for example, to understand the concentrations of chemicals in rivers or lakes and their effects upon a community, ecologists may need to have general and special knowledge of some aspects of chemistry and of social sciences. If called upon to prepare environmental impact statements, ecologists will need all of these plus communication skills. The Ecological Society of America has stated that the demand for ecologists will continue as long as environmental problems and issues persist.

Although some ecologists are engaged in research, for many teaching is a basic function. Private industry and government use many ecologists as consultants. Consulting may involve practical application of the ecologist's special skills, involving interactions with people of many professions, including economists, lawyers, planning personnel, as well as other scientists. Biologists who become interested in ecology as a profession should have a thorough background in biology. A study of wildlife biology will usually be helpful. In addition to the usual courses in chemistry, physics, and mathematics that all biology majors should have, a study of soil science, climatology, basic statistics, computer science, and geology is recommended.

Teachers who are oriented toward ecology may find employment in secondary schools as well as collegiate institutions. University departments that hire ecologists include the usual bioscience departments—biology, botany, zoology, and microbiology—and also specialized departments, such as forestry, fisheries, biology, horticulture, agronomy, entomology, oceanography, and wildlife management; there are also interdisciplinary programs. In the federal government, the Forest Service, Soil Conservation Service, National Park Service, Fish and Wildlife Service, and U.S. Public Health Service are examples of agencies that utilize the skills of ecologists. Environmental organizations such as The Sierra Club, National Audubon Society, and Greenpeace rely on the expertise and activism of ecologists. The manufacturers of agricultural products, especially fertilizers and insecticides, hire ecologically oriented personnel, as do the members of the paper and oil industries where pollution is a problem. Several agencies of the United Nations have employed ecologists to gather data relating to the human environment in many parts of the world. Others have served as advisers.

From the above discussion, it should be clear that ecologists are vitally interested in conservation. Special careers in wildlife conservation are available. The wildlife biologist aids in the management of game, fish, fowl, fur-bearing animals, or any type of wildlife. The biologist is interested in the habitats of all these animals and may help in the preservation or restoration of streams, lakes, or marshes. Sometimes the work takes the wildlife biologist into remote areas, often under rugged and difficult field conditions. Other wildlife biologists may work in laboratories, concentrate on teaching, take part in extension work, or write or give lectures in their specialty. The research that wildlife biologists do may be related to taxonomy, physiology, genetics, and many other fields as applied to wild animals, fish, or birds. The research may be of a basic scientific nature or have practical applications.

Most of the work of the wildlife biologist will be directed toward learning as much as possible about wild animals so as to preserve their existence and to lessen the impact of our civilization upon them. The biologist who chooses to devote his or her time and talents to ecology, conservation, and wildlife biology will be rewarded not only by a fascinating study, but by the knowledge that his or her future and that of humankind are intimately bound. Ecologists and wildlife biologists are prominently involved in studies on the environmental aspects of the search for new sources of energy. Among their concerns are aid in the proper design, location, and operating methods of power plants, the effects of high voltage on wildlife, and the development of old mining land into a suitable habitat for wildlife. Whether you are of an idealistic or a practical nature, you may find these challenges to be rewarding.

CHAPTER 5

CAREERS IN THE BIOMEDICAL SCIENCES

Biologists who choose to work in one of the biomedical sciences find themselves at the very forefront of advances in knowledge of health and disease, in factors governing aging, and in improvements in the physical and mental well-being of humanity. Nearly all medical advances have been based upon observations previously made in the laboratory of the bioscientist. The discoveries made in these laboratories have applications not only in medicine, but in many other fields of benefit to society.

PHYSIOLOGY

In previous chapters we have pointed out that biology is not merely one science but a group of sciences. No branch of biology illustrates this better than does physiology, which offers a rich variety of careers. Physiology is placed at this point in our discussion largely for the convenience of the author. You should understand that physiology is a broad term, embracing the interests of all life scientists whose primary concern is with the processes that sustain life. The present discussion will be confined to animal physiology (which includes, of course, human physiology); plant physiology has been considered in the preceding chapter.

The interests of physiologists are so varied that it is difficult to define physiology, and there is no clear-cut line separating it from closely related biosciences. It is usual to consider physiology as being the science concerned with the functions and operations of a living being or any of its parts, and of the interrelationship of one living part to another. One leading physiologist has said, "Physiology...has contributed more than any other science to our present knowledge of the nature of man." Another has said that the "physiological sciences...have the greatest relevance of any field of science to the human condition." Such statements are more than proud boasting. In a sense, all the biomedical sciences to be discussed in this chapter may be considered as branches of physiology, or they developed into independent disciplines from physiological origins but can still be considered to be part of the general category of physiological sciences.

Physiologists are concerned with such matters as: What makes animals (or people) grow? What regulates the rate at which they grow? How is food digested? What regulates the activities of the digestive tract, and how can these activities be modified? How are the products of digestion absorbed into the bloodstream, and then how are they distributed to places where they are needed? Why do we need blood, and what governs its circulation? What are the intimate mechanisms involved in reproduction for the perpetuation of the species? How do people respond to changes in their environment? What are the physiological effects of pollution? What governs our responses to the day/night cycle or to the annual changes of the season? How do we respond to changes in air pressure such as those encountered in mountain climbing or in high-altitude aviation?

These are only a few of the questions engaging the attention of physiologists; the list could go on and on. To find the answers, physiologists study living creatures, including people themselves. The results of such inquiries form the very basic structure of medicine, whether human or veterinary. Scientists in related disciplines—such as biochemistry, pharmacology, biophysics, and psychology—must have a background of knowledge of physiology. Although the information gained by the study of animals most closely resembling human beings—that is, the mammals—has the most obvious application to human physiology, much has been learned from observations of creatures far removed from the human on the evolutionary scale.

A few examples will illustrate some of the things that have interested physiologists.

- One physiologist studied for many years the water balance in frogs; that is, the intake and output of water by the body of the animal.

One might inquire what the practical use of such an activity could be. To the scientist, such a question may seem irrelevant, and the experiments may have been done simply as a response to the need of the human mind *to know*. But in this instance, the simple experiments in the frogs were destined to lead in many directions. Other physiologists took up the study and discovered that the electrical and chemical forces involved in transporting water and salt through frog skin could also be involved in the functioning of nerve cells and the kidneys in human beings. These investigations, in turn, were eventually, after many further steps, put to practical use by physicians wishing to regulate kidney output in patients with certain kinds of heart disease. But that is not all. The knowledge gained in these studies has also proved to be invaluable in aiding people to cope with desert conditions; much use was made of it in the desert campaigns in North Africa during World War II.

- The migration and breeding cycles of migratory birds have long excited people's curiosity, and they still present many unsolved problems. It has been found that one factor in some of the cyclic behavior of birds is the length of day. Light, and a specific number of hours of light during a single day, perceived by the visual apparatus of the bird, triggers the release of a chemical substance (a hormone) in a portion of the bird's brain called the hypothalamus. Cyclic behavior related to the time of year, and hence to length of day, is also characteristic of many other animals, including some mammals. No one seems to know, however, whether a similar factor may be involved in the release of hormones in human beings. Might this give a rational basis for the old adage, "In spring a young man's fancy turns to thoughts of love"? Do we interfere with the natural cycle of hormonal events by artificial lighting? We know that it is so in chickens, and poultry farmers take advantage of it.

- A physiologist studying the flight muscles of a house fly might be thought to be engaged in a line of research having little relevance to human medicine. However, a substance has been found in these tiny organs that is correlated with aging in the insect. With more and more people living into the retirement years, the processes involved in aging become of increasing importance. No one can predict to what heights basic discoveries may lead, no matter how humble a creature may be the subject of inquiry or how irrelevant the question under investigation may seem on the surface to be. This is a point to keep in mind the next time you read of a scientist's investigation being held up for ridicule by someone ignorant of science just because the title of the research does not have any obvious application to human problems.

Although there are some general physiologists, most physiologists specialize, and the list of specialties is a long one. The cellular physiologist studies processes that go on within the microscopic world of individual cells. The cells may range all the way from bacterial cells to those derived from human tissues, and information derived from one form often (though not always) applies to many others. In this way, laws are found that govern vital processes.

Other physiologists may become engrossed in the functioning of some particular organ of the body or a system of organs. For example, a physiologist who studies the endocrine system is an endocrinologist; her or his specialty is the system of glands that release secretions into the bloodstream to be carried to other organs and assist in the regulation of growth, metabolism, reproduction, and many other processes. Even an endocrinologist may specialize and may concentrate on, for example, the reproductive system. Some physiologists study particular groups of animals such as insects, fishes, or even protozoa. The comparative physiologist is interested in those aspects in which one form differs from another; these findings are especially valuable to the veterinary physician as well as to the stock breeder or the pet fancier. You might find an exciting challenge as an environmental physiologist who can make direct and immediate contributions to our knowledge and awareness of the effect of society's changes in the environment, as well as of exposure to unusual stresses, such as certain tropical regions, the arctic, high altitudes, or space. It is not my intention to list all the kinds of physiologists there are, but merely to indicate by a few examples the surprising diversity available to you if you are considering physiological science.

The level of education you need as a physiologist depends largely upon what position in the physiological team you may assume. There are many openings for laboratory assistants and technicians; these will be discussed in a later chapter. Such positions may sometimes be entered directly from high school or perhaps with an associate in arts degree from a two-year college. However, for a more advanced position in physiology, a more prolonged education is necessary, and the top-notch jobs in physiology are held by those with Ph.D. or M.D. degrees. Colleges and universities employ about two-thirds of all physiologists. They are engaged in teaching, research, or both. Not all colleges have departments of physiology; at the undergraduate level of teaching, most physiologists have appointments in such departments as biology, zoology, and entomology. Agricultural, dental, and veterinary schools also employ physiologists. Medical schools are especially important, for physiology is the very foundation of medicine, and physiologists are employed not only in departments of physiology and other basic sciences, but often in clinical departments—medicine, surgery—as well. Clinics, hospitals, private research foundations, drug companies, U.S. government laboratories and regulatory agencies—these and many others employ physiologists.

From time to time, the American Physiological Society surveys the educational institutions of the United States, Canada, and Mexico and lists all those offering degrees in physiology. It should be emphasized, however, that the future physiologist's choice of a college is by no means limited to one that offers an undergraduate major in physiology. Indeed, the majority of physiologists now at the height of their careers have specialized in physiology only in graduate school. A good background in biological science, however, is essential. This will include both general and advanced biology, physics, and general, organic, analytical, and physical chemistry. In recent years, the importance of mathematics has increased rapidly. Even laboratory assistants may need algebra in helping to plan experiments and analyze data, while the professional physiologist will need to understand calculus and probability theory not only to plan and interpret her or his own work, but to comprehend the work of others. If all of these courses are completed in the undergraduate years, the student will have a good background for specializing in physiology in graduate school. The student physiologist also should acquire some knowledge of related biomedical sciences, especially biochemistry and anatomy.

The boundary between physiology and some of the other biosciences is indistinct; indeed, the overlap is so great that an individual trained in one life science sometimes ends up working in another. For example, nobody can tell where physiology lets off and biochemistry begins. Physiology also overlaps with pharmacology, anatomy, genetics, embryology, parasitology, ecology, psychology, and other disciplines. For the convenience of the author as well as that of the reader, there must be some sort of classification; therefore, we shall proceed with a consideration of some related fields.

BIOPHYSICS

Biophysics is one of the newest of the biological sciences, but it has grown rapidly. It is so closely related to physiology that in some schools biophysics and physiology are combined into one department. However, it has become sufficiently independent as a discipline for biophysicists to have organized their own society, the *Biophysical Society.*

The biophysicist's approach to biological problems utilizes the methods of modern physics, higher mathematics, and physical chemistry. The biophysicist addresses some very basic issues, and the findings often have practical consequences. Biophysics has contributed to the study of cancer; to the design of computers, which are analogous in some respects to the brain; and to the development of new drugs. Many biological problems cannot be approached without the aid of the instruments and methods of the physical sciences and even with the physical scientist's point of view.

All matter, living and nonliving, is made up of molecules, and the most complicated molecules in our bodies are the proteins. A protein molecule is made up of thousands of atoms, each arranged in precise order and spatial relation to each other. Most of the hormones of the body are proteins, as are the enzymes—those important "ferments" that enable our bodies to carry on complicated chemical reactions far beyond the abilities of the most brilliant chemist to duplicate in the test tube by usual chemical means. Biophysicists have contributed greatly to understanding the reactions involving enzymes. The outstanding advances in biotechnology, genetics, and biomedical engineering discussed in Chapter 4 were all made possible by the contributions of the biophysicist. Among these was the success in synthesizing the molecule of an enzyme associated with reactions involving RNA (ribonucleic acid).

Many of the reactions within the cells of the body are those involving large and complicated protein molecules, which include among other things reactions in the immune mechanisms, that is, those mechanisms involved in defenses against foreign substances. The biomedical engineer has had some success in prolonging people's lives by replacing failing organs with transplants of healthy ones. These operations have been much heralded when they are successful, but there have been instances of disappointment and failure, often because of harmful interactions between the proteins of the transplanted organ and those of the patient receiving the transplant. In other words, the body of the recipient treats the transplant as a foreign and hostile intruder. There has been some success in avoidance of such interactions, but there remains a large problem. Increased knowledge of interactions between proteins offers the most promising approach to the solution of such problems and constitutes an important challenge that future biophysicists can help to meet.

Other interests of biophysicists include the effects of radiation on living matter, the details of the way in which the retina of the eye converts the energy of light into a signal to the brain, and the biological effects of gravity, of laser beams, and many others. Biophysicists have also contributed to the design and functioning of cardiac pacemakers—devices that initiate heartbeats in persons whose own natural pacemakers have ceased to function properly. All interactions of energy with living matter, whether animal or plant, interest the biophysicist. Hence, biophysicists seek to learn how plants convert the energy of sunlight into chemical energy that can be stored in the plant in the form of carbohydrate, which in turn supplies energy to our own bodies. Responses of plants to gravity (stems grow up and roots grow down) and the turning of plants toward a source of light are other examples of problems of the biophysicist whose answers might be not only of scientific interest but of practical importance.

From this brief description of biophysics, you can see that a biophysicist is a physiologist with a very special approach to physiological problems. The student aiming for a career in biophysics will need an extraordinarily broad base in physics, chemistry, mathematics, and biology, studying these disciplines in high school and continuing them through college and beyond to the doctorate. Some biophysicists recommend as many as ten courses in college mathematics. The student of biophysics will become familiar with the concepts of such subjects as protein chemistry, quantum mechanics, information theory, solid-state physics, and probability theory.

Although the total number of openings for biophysicists may not be as great as for some other disciplines, the high degree of training required limits the number of competitors for the jobs available. Most biophysicists are employed by universities in teaching and research.

Some are employed in space installations and in the armed forces. Large hospitals, laboratories of nonprofit research foundations, and drug companies also make use of the talents of biophysicists.

Biophysics is an intellectually demanding discipline to be undertaken only by those who are willing to work hard and who enjoy mental stimulation. It is a highly experimental field, challenging and exciting, and utilizes the most modern concepts and instruments. If you have the basic motivation for such a study and the capacity to meet the challenge, you will find it a most rewarding vocation.

BIOCHEMISTRY

As the name implies, biochemistry is both a biological science and a branch of chemistry. It may be taught in departments of biology, physiology, or chemistry, but many universities have independent departments of biochemistry. Biochemistry is so closely allied with physiology that it would be difficult to make a statement about the difference between them without saying something that would bring an objection from either the physiologists or the biochemists. Perhaps it is fair to say that in most cases the difference is one of emphasis. Although the physiologist may use biochemical techniques to elucidate the function of a tissue or an organ, the biochemist will take special pleasure in studying the chemical processes that occur in active tissues. Both physiology and biochemistry evolved from general biology, but biochemistry could not arise until the advent of organic chemistry.

Considerable progress was made in physiology during the nineteenth century, but it was not until near the close of that century that chemists became aware that all organic substances contain carbon. The chemistry of carbon compounds then became organic chemistry. Biochemistry emerged as a separate discipline when some scientists combined biology and organic chemistry and began to study chemical reactions occurring in living systems. Biochemistry is hence a younger science than physiology and in a sense, may be regarded as the offspring of a marriage between physiology and chemistry. The lineage of biochemistry is apparent in the various names that are sometimes given to it, such as physiological chemistry and biological chemistry.

Biochemists spend their time in many different ways. Those with the highest degrees of education are usually involved in teaching or research, although a growing number may be in administration or in service work. Young people in the early stages of their careers are widely employed as laboratory workers, technicians, or assistants. Many find this type of work to be fully satisfying; advancement as a fully qualified professional research scientist, for example, does not appeal to everyone.

Testing and analytical programs occupy the time of many biochemists, while others spend full-time on research and still more are teachers; most teachers in colleges or universities devote part of their time to research. The research may be basic, that is, performed primarily to increase scientific knowledge; or applied, that is, to solve problems with immediate practical application. The research may be directed to such questions as: How do living things obtain energy from food? How do green plants convert carbon dioxide and water into carbohydrates with the aid of sunlight? How can we learn to imitate this process in the laboratory? What is the chemical basis of heredity? What biochemical processes are involved in memory, learning, and thinking? How do hormones work? What causes abnormalities? What fundamental changes occur in disease? What are the biochemical accompaniments of death? Can the changes in disease or death be modified so that health can be restored?

These are only samples of the many questions that biochemical research may seek to solve; there is scarcely any limit to the number of such questions that might be addressed. Many times the research biochemist or the clinical chemist must apply newly discovered techniques to her or his work. This may involve the use of new instruments or new chemical reagents. If existing methods and procedures do not fully meet requirements, the biochemist will need to invent them, test them fully in the laboratory, then publish the results so that other biochemists may confirm them. In addition to research and teaching, some biochemists, especially in hospital laboratories or in independent testing laboratories, provide services that assist physicians in diagnosing and treating disease. Others, especially those employed by industry, also may aid in quality control.

Colleges of pharmacy, dentistry, medicine, veterinary medicine, and agriculture are among the professional schools that provide opportunities for teaching or research in biochemistry. In these schools, it is not only the departments of biochemistry that employ biochemists; employment also is offered in departments of physiology, pharmacology, microbiology, anatomy, medicine, and others. Departments of biology or chemistry in colleges and universities supply positions for many biochemists.

Hospitals, independent research institutes, public health departments, government laboratories such as the famed National Institutes of Health, and agricultural experiment stations are among those that benefit from the special talents of biochemists. A wide variety of industrial laboratories, including food processors, drug and cosmetic manufacturers, and the huge chemical and petroleum industries, constitute a continuing source of employment opportunities for those interested in applied biochemistry, in basic or applied research, or in scientific administration. Many biochemists, especially clinical chemists, have gone into business for themselves, offering biochemical testing services on a fee or a contract basis. Such testing laboratories may employ numbers of biochemists, as well as biochemical laboratory assistants and technicians. The workload of testing laboratories, including those owned and operated by hospitals, has a record of growth; it has been said that the load doubles about each seven years. This, then, constitutes a continually growing demand for those prepared to meet it.

The education needed by a biochemist will depend upon what position on the ladder the person wishes to take. For advancement to the top ranks, advanced degrees are desirable, and for the top-notch jobs, only a doctorate will suffice. Training leading to such degrees may usually be obtained while working as a laboratory assistant or teaching fellow. The education of the biochemist need not differ much from that of other specialists in the health sciences, but a good grasp of chemistry, mathematics, and physics is of special importance. Generally speaking, it is not necessary to major in biochemistry during the undergraduate years; most biochemists have majored either in biology or in chemistry, then concentrated on biochemistry in graduate school. Nearly two hundred colleges and universities offer graduate degrees in biochemistry.

Some biochemists have become interested in their specialty after attending medical school and receiving either an M.D. or Ph.D. degree. Biochemists with a medical background usually are employed in medical schools, hospitals, and research laboratories. By whatever route the biochemist enters the profession, he or she will find many opportunities for creative and exciting work.

PHARMACOLOGY

Although pharmacology is a basic science that is an important part of the training of all students in all health professions, including pharmacy, medicine, dentistry, and veterinary medicine, the word *pharmacology* is still not understood by many. Only within recent years has there been a move toward offering pharmacology at the undergraduate level in colleges of liberal arts, and the majority of schools do not have such programs. Most medical schools and many pharmacy schools offer courses leading to master's or doctor's degrees in pharmacology. The students who are accepted as candidates in these programs are usually graduates of colleges where they have majored in one of the biological sciences. The biology major should, therefore, become acquainted with the career potentialities offered by pharmacology. Young people trained in this discipline have had little difficulty getting jobs, even during periods of recession.

To define pharmacology as briefly as possible, it is the study of the interactions of drugs with living systems. By *drug* we mean any chemical substance that can be given to a human being or animal that will affect the recipient for good or ill or that will in any way affect the life processes. The living system might be a microscopic part of a cell, an organ of the body, or a whole animal or person, healthy or sick. Physiologists also study responses of cells, organs, and organisms, and biochemists study their chemical reactions. Pathologists examine the effects of disease. Pharmacology, then, overlaps with those disciplines and uses their techniques. No one could be a top-notch pharmacologist without a thorough background in physiology and biochemistry.

Physicians and pharmacists also know a great deal about drugs, but to be a physician or a pharmacist is not necessarily to be a pharmacologist. The pharmacologist's knowledge of drugs is unique. The intimate nature of drug action as affected by the size of the dose, method of administration, solubility in various body fluids, the effects of small changes in chemical structure of the drug, and the significance of physical properties as related to drug action are among the matters of concern to the pharmacologist but outside the specific interests of physiologists or biochemists.

Pharmacology is divided into many branches as are most other life sciences. The general pharmacologist tries to understand the action of a drug in all parts of the body and often is searching for better and safer drugs for treating disease. The interests and methods of comparative pharmacologists might be similar, and their findings might be of interest, for example, to a veterinary physician who needs to treat various species of animals. Molecular pharmacology is more basic in its approach and seeks to discern the interaction between a molecule of a drug and a molecule within a cell of the body. Some of the discoveries of the molecular pharmacologists have given physicians valuable information about such things as the intimate mechanisms of the nervous system or the actions of the kidneys. Biochemical pharmacologists or biophysical pharmacologists use the techniques of physics and biochemistry; sometimes their primary interest is to determine what the body does to the drug rather than what the drug does to the body. Such information is especially valuable in helping to improve the drugs already available.

Toxicology is a branch of pharmacology specifically concerned with the adverse effects of chemical substances, ranging all the way from minor unwanted effects of useful drugs to the actions of virulent poisons. The toxicologist is equally interested in the immediate effects of a single exposure to a substance and the long-range actions of a drug that a person may need to take daily for many years. Toxicologists are in great demand in the drug industry. No drug is 100 percent safe; one of the duties of the medical or forensic toxicologist is to determine the limits of safety of a drug. One would hope to find out enough so that we could never again have a tragic occurrence such as the armless babies who were born of mothers taking thalidomide. Forensic toxicologists are also involved in research into the medical and legal aspects of drug abuse. The industrial toxicologist applies his or her knowledge to the health of workers in industry; a wide variety of industrial concerns utilize such expertise. The environmental toxicologist helps protect public health by studying the effects of water and air pollutants, industrial wastes, household chemicals, food additives, pesticides, and the many other exposures of the modern world.

Some pharmacologists concentrate their studies upon some particular organ or system in the body. For example, the action of drugs upon the nervous system comes under the attention of neuropharmacologists. Their interests are related to those of behavioral pharmacologists who use the techniques of psychology to ascertain the effects of chemical substances upon behavior and psychic functions. These branches of pharmacology have become especially important since the use of drugs that alter behavior has become so widespread.

Deserving of special attention is the clinical pharmacologist who applies the discoveries of other pharmacologists to the human subject and who, of course, also makes many discoveries. Clinical pharmacologists seek to determine whether a particular drug is useful, how it should be used, and what adverse effects it may have. They bridge the gap between the original discovery of a new drug and its use by a practicing physician. They must, therefore, be versed not only in physiology, biochemistry, and pharmacology but in medicine as well. They possess M.D. degrees and sometimes Ph.D.s also. The training of a clinical pharmacologist is a long one, and this specialty is one of the most important and demanding of all the health sciences.

Employment opportunities in pharmacology are available in medical, pharmacy, and veterinary schools—and to an increasing extent in four-year colleges. Pharmacologists are also in demand in government laboratories, especially such agencies as the National Institutes of Health and the Food and Drug Administration. Hospitals and independent nonprofit research laboratories hire many pharmacologists, as do the laboratories of industry, especially the pharmaceutical industry. Several independent laboratories engaged in providing service to other institutions under contract also hire pharmacologists. The variety of activities in which pharmacologists are engaged matches the variety of institutions for which they work. Some spend most of their time teaching, some are in basic research, some are engaged in trying to discover new and better drugs for the treatment of human or animal diseases, and others help in the protection of public health and safety.

There are many opportunities for working in a pharmacology laboratory for those without complete professional training, but for the top-level jobs a doctoral degree is essential. During the high school and undergraduate college years, the education of a future pharmacologist parallels that of a physiologist or biochemist. Most helpful will be courses in mathematics through calculus, including statistics; a course in general physics; and chemistry, including organic and physical chemistry. With such a background, whether or not the curriculum includes an undergraduate course in pharmacology, the student will be well equipped to pursue graduate study in most of the basic medical sciences and need not make a final choice among them until graduate school.

NUTRITION

There are several biosciences concerned with food and nutrition in one way or another. We have already seen that physiology and biochemistry may deal with digestion and assimilation of food and with the distribution and disposition of the products of digestion after they are absorbed into the bloodstream. In the next chapter, we shall comment on careers in food technology. Nutrition overlaps those disciplines, but it may also be considered a bioscience in its own right. The dietician or nutritionist studies the fate of foodstuffs after digestion and absorption (thus overlapping with physiology and biochemistry) and is especially concerned with the relationship of food to the health and welfare of both the individual and society, and its connection with the cause, treatment, or prevention of disease.

There should be no doubt about the importance of a knowledge of nutrition so long as we live in a world in which much of the population goes to bed every night with unfilled or partially filled stomachs. We have all heard of the catastrophic effects on population in the past by diseases having their roots in nutritional deficiencies; for example, the huge losses of manpower from scurvy during the days of the old sailing vessels before ships were provisioned with lemon juice. Scurvy was mainly caused by a lack of vitamin C; but it would be incorrect to think that deficiency diseases are a thing of the past. An example is kwashiorkor caused by a protein deficiency and prevalent today in much of the underdeveloped world. The victims are mainly children between the ages of six months and three years who simply cannot eat enough to fulfill their protein requirement with the type of diet that prevails where they live. Nutritional deficiencies also occur in impoverished areas, both urban and rural, in our own country.

Many groups, such as the Food and Agricultural Organization of the United Nations, as well as individual nations, industries, and independent research organizations, are seeking ways to correct these conditions. Those with training and skill in the science of nutrition will find useful and rewarding work in these areas. In our own society, although we know how to prevent such diseases as scurvy, pellagra, and beriberi, research is still underway on the nutritional aspects of hypertension, heart ailments, obesity, tooth decay, and possibly even arthritis and cancer. Some unfortunate children are born with errors in metabolism of foodstuffs that lead to mental retardation and early death; an example of this is the disease known as phenylketonuria. Largely because of the efforts of nutritionists, many of these children have been able to develop in a nearly normal fashion, and research is continuing.

Some nutritional discoveries have far-reaching results. I can remember when a diagnosis of pernicious anemia was almost the same as a death warrant. With the discovery of vitamin B_{12}, the means for treating this disease was at hand. But B_{12} was to have an unexpected and quite unrelated significance in nutrition, for it proved to be a factor that increased the efficiency with which meat animals converted hay and grain into body-building protein.

There are several types of careers related to nutrition. Research dieticians are scientists contributing new knowledge on such subjects as human or animal metabolism, the interrelationships of foods, and the effects of nutrients on health. They are employed in universities; in medical, dental, veterinary, and agricultural colleges; in government agencies; in international agencies; and in industry, especially industrial concerns engaged in supplying dietary supplements for either human or animal use. Some research dieticians work for food manufacturers or grocery store chains, analyzing the nutritional content of food for labeling purposes or for marketing. Community dieticians apply the science of nutrition in various fields of health care and are primarily employed by home health agencies, health maintenance organizations (HMOs), and human service agencies. Administrative dieticians apply the principles of nutrition in the planning and management of food service programs in hospitals, schools, and other institutions. Some dieticians prescribe diets, select foods, and counsel hospitals, clinics, nursing homes, and other individuals or communities; they are clinical dieticians.

The wide range of different kinds of jobs in nutrition is reflected in the range of educational requirements. The American Dietetic Association maintains a list of colleges that provide training courses for high school graduates to qualify as dietetic assistants or dietetic technicians; these courses may lead to an associate degree. There are also undergraduate programs leading to a baccalaureate degree as a dietician; other graduates may become approved dieticians after a suitable internship. In many states, there are universities offering advanced training to the master's and doctoral levels also. The education and training of the research dietician is comparable to that of other bioscientists. Physiology, microbiology, organic chemistry, and biochemistry are essential. Mathematics, statistics, computer science, psychology, sociology, and economics should not be neglected. As in all sciences, the research nutritionist with a doctoral degree has an advantage.

IMMUNOLOGY

Immunology is the study of the way in which the body protects itself against a foreign invader, whether it be a substance producing an allergic reaction, an invasion by a disease-producing microbe, an organ transplant, a parasite, or a virus. The reaction of the body to each of these insults is called an immune response. Each immune response is very specific; that is, each particular kind of invader produces an immune response that is specifically directed toward that invading material. The invading stimulus is a chemical substance (called an antigen) and the body produces another chemical substance (antibody) that reacts chemically with it. Hence, immunology could be considered from one point of view to be a branch of biochemistry. Since the immune system is one of the normally functioning systems of the body, immunology is also a branch of physiology. Since an immune response is often a part of a disease process, immunology is to some degree a part of the science of pathology. And since immune responses sometimes have to be treated by physicians, immunology is also a part of medicine.

Immunology, then, is another example of the overlap among biomedical sciences that we have mentioned in previous pages. It is, however, a science in its own right, and professional immunologists have their own society, the American Association of Immunologists. Some of the triumphs of immunological research have been of inestimable value to society, and the potentialities for future developments form an exciting prospect. No doubt you have heard the story of the eighteenth century English physician, Edward Jenner, who noticed that dairy workers who had had cowpox—a relatively mild disease—did not thereafter get smallpox—a serious and often fatal disease that was prevalent at that time. Jenner was then able to protect others by deliberately infecting them with cowpox.

Also familiar is the story of the great French bacteriologist Louis Pasteur, who through an oversight (even the greatest of scientists make mistakes) left a dish on the laboratory shelf while he went on vacation. The dish contained bacteria that could cause cholera in chickens. But the conditions to which the bacteria were exposed rendered them incapable of causing disease. Instead of dumping them down the sink, Pasteur experimented further with them and found that chickens receiving them were protected against cholera. From this beginning, he made many other vaccines. Today we have many vaccines, useful against many diseases that formerly killed millions of people each year but that now are seldom seen in most parts of the world. The ability of a great scientist to turn a mistake into a brilliant discovery has been shown more than once in the history of bioscience—the story of Alexander Fleming and the discovery of penicillin has been mentioned in a previous chapter.

Diseases caused by bacteria and viruses are not the only things that excite the interest of immunologists. If you suffer from hay fever, if you have symptoms like a common cold when you breathe house dust or animal danders, if you have a severe rash after contact with poison ivy, if insect stings cause severe reactions, then you are experiencing an immune response that, instead of protecting you, is causing you distress. Some people even develop such responses to some of their own cells or tissues—these are called autoimmune disease; perhaps the most common of these is rheumatoid arthritis, which makes millions of people suffer. All of these furnish challenges to the immunologist, who seeks knowledge that will enable physicians to alleviate the symptoms of such disease—and in the end, cure or prevent them.

Other challenges for the future of immunology include the control of adverse reactions in blood transfusions and in the transplantation of tissues and organs. Of the many thousands of kidney transplants that have been made, hundreds have been failures, and immune responses have been the cause of as many failures as all other causes combined. Especially exciting are the prospects for possible immunological treatment of cancer, or for the selective removal from the blood of troublesome substances that trigger immune responses, or for reducing the risks from those responses that are inherited. Finding a cure for the deadly disease AIDS also presents a challenge for immunologists. No wonder that the National Institute of Allergy and Infectious Diseases of the National Institutes of Health has referred to immunology as "a discipline whose time has come."

Recent advances in biotechnology have given immunologists new tools to use in their research. Substances known as monoclonal antibodies, are produced by hybridoma, a special type of cell created by fusing two different kinds of cells. The interests of immunologists thus overlap those of geneticists, microbiologists, and bioengineers, and any distinction between immunology as a basic science and as a practical discipline must be arbitrary—as indeed is the case with all the biomedical sciences.

Immunology presents career opportunities ranging all the way from laboratory assistants and technicians to professional scientists in charge of large laboratories. At all levels, the work is of absorbing interest and gives great personal satisfaction. The worker in immunology uses many of the laboratory techniques of biochemistry and microbiology. The better the person's preparation in these fields, the better he or she will be able to perform the duties, and the greater pleasure the worker will derive from the daily activities. For full professional standing and the best-rated jobs, the undergraduate preparation should be about the same as we have described for the other biomedical sciences—namely, a grasp of biology, including microbiology, with chemistry through organic and preferably physical chemistry, a course in physics, some knowledge of computer science, and, of course, a basic understanding of mathematics.

Immunologists employed by educational institutions divide their time between teaching and research; this includes more than half of all members of the profession. Training in immunology may be obtained in almost any university having a medical school, and there are a few immunologists in departments of biology. There are relatively few departments of immunology; most often it is a section within a department. Departments of microbiology, biochemistry, and medicine most frequently employ immunologists. A few hospitals and research foundations also give employment to immunologists, but the leading employer outside the medical schools is the federal government, especially the National Institutes of Health. There are also some openings in laboratories supported by local and state governments.

In opportunities to be on the frontier of science, no discipline surpasses immunology.

PATHOLOGY

Pathology is both a basic biomedical science and a specialty of medicine. By the use of the techniques of biology, chemistry, and physics, pathologists examine the tissues and fluids of the body. They seek to determine the presence of any measurable or visible changes produced by disease or by any other interference with normal bodily structure or function. Plant pathologists were discussed in Chapter 4; in this section, we shall be concerned with those pathologists who examine specimens of human or animal origin. Their findings enable them to judge whether a given specimen has come from a healthy individual or if there is disease or injury. The nature of a disease and the effectiveness of treatment may also be revealed.

Pathology, then, is a link between the basic biomedical sciences and clinical medicine. Most pathologists are physicians who have specialized in this particular field. Medicine and its specialties are beyond the scope of this book; we are interested in pathology as a biological science. And, in fact, there is a demand for pathologists who are not physicians. They include dental and veterinary pathologists as well as a growing number of individuals in research and service jobs who have a Ph.D. in pathology. Indeed many of the members of the American Society for Experimental Pathology have advanced work in areas other than that for the M.D. degree.

In addition to the top-grade professional jobs, there is a demand for workers in pathology laboratories especially trained for the many demanding technical jobs. These laboratory technicians and medical technologists will be discussed in the next chapter, but we should mention at this point that the demand for such workers is very great. Physicians are placing more and more reliance on the findings of the laboratory to assist them in diagnosing disease and in following the progress of therapy. Some of the laboratory jobs also give an opportunity for the young worker to study for the most advanced professional qualifications.

Some diseases, such as diabetes and leukemia, can be identified only by laboratory means. The pathologist is essential for the diagnosis of such diseases and supplies the means for following responses of the patient to treatment. For supervising a laboratory bearing such important responsibilities, a pathologist should qualify for a special certificate from the American Board of Pathology. In most cases, the pathologist also will be a teacher, aiding in the training of medical students, nurses, interns, medical technologists, and resident physicians.

Teaching, research, and service—these are the outlets for the pathologist's skills. Nearly every accredited hospital employs pathologists. A pathology department is a prominent feature in medical schools; dental and veterinary schools also have such departments. Some pathologists operate independent laboratories that provide service to physicians in private practice. Independent research institutions operated by nonprofit foundations employ many pathologists and pathology technologists. A particularly important source of employment opportunities is the federal government, as well as state and local public health installations. The armed services operate large institutes for research and service in pathology, and the National Institutes of Health and the Food and Drug Administration have particular need for those with skills in pathology.

Laboratories concerned with studies on the safety of drugs provide a large and expanding demand. Individuals with Ph.D. degrees in pathology, or with training in veterinary pathology, are especially in demand in these establishments. Studies on the safety of new drugs must be conducted at length and with great skill on experimental animals before the drug can be tried on human beings. After the animals have received the new drug, the fluids and tissues of their bodies must be subjected to the same tests as those made on human patients in disease. In fact, no patient ever gets quite as thorough an examination as do these animals. Federal laws and regulations make such studies mandatory, and the standards are constantly being elevated; hence, the demand for personnel is very great.

The tests just described are performed in the laboratories of drug manufacturers and in many laboratories operated by the government, such as the National Institutes of Health and the Food and Drug Administration. In addition, many independent laboratories carry out similar studies. Several commercial pathology laboratories perform such tests for others on a contract basis.

CHAPTER 6

OTHER CAREER OPPORTUNITIES IN APPLIED BIOLOGY

THE AGRICULTURAL SCIENCES

The application of biological and physical sciences to the production of food and fiber forms the very foundation of the success of our civilization. It has helped to make American output levels the envy of the world. It is no wonder, then, that an extraordinarily large and varied number of careers can be found in these fields. The *Occupational Outlook Handbook* of the Bureau of Labor Statistics, U.S. Department of Labor, groups agricultural scientists, like biological scientists, in the category of life scientists. It estimates that in the mid- to late 1990s, agricultural scientists held more than 25,000 jobs. An additional 18,000 people were employed in agricultural science faculty positions in colleges and universities.

Agricultural scientists come from a wide variety of training programs and engage in many different kinds of work. Some of them have already been discussed; such disciplines as plant pathology, entomology, ecology, fisheries biology, and genetics are among those described in previous chapters that have applications here. Examples of agricultural sciences that may be regarded as applied biology include agronomy, which applied biological science to such practices as crop breeding and production, the testing of new varieties of plants, and plant propagation. Horticulture is a related field, embracing not only breeding and culture of plants, but problems in storing and handling. Ornamental as well as food plants are considered. Forestry is a related discipline; more about forestry will be given in a following section of this chapter.

Animal husbandry is an agricultural science concerned with the breeding, nutrition, and overall quality of livestock. Poultry science and dairy science are sometimes considered as separate branches of animal husbandry.

Soil scientists are interested in the biological effects of soils as affected by physical, chemical, and other properties. Food technology applies science to the production, processing, preparation, and distribution of food. For example, such familiar items as freeze-dried coffee, frozen orange juice concentrate, and dehydrated vegetables are the results of the efforts of food technologists—as are the packaging innovations that permit them to be marketed. The food technologist must know something of nutrition, physiology, biochemistry, microbiology, and related sciences.

A highly developed specialty is that of seed technologist. Seed technology, a surprisingly wide-ranging field, has a place for people trained in botany, genetics, agronomy, entomology, and several other specialties. The seed technologist and seed analyst may supervise harvesting procedures of seeds, conduct tests for purity and germination, evaluate storage procedures, devise methods for control of insects and fungal infestations, and perform many other procedures. Research in the field may be concerned with breeding improved varieties, increasing seed yield, and improving processes involved in the handling and distribution of seeds. Special problems and opportunities may be presented by the introduction of new varieties resulting from genetic research, and the seed technologist keeps abreast of these developments.

It has been estimated that there are about 350,000 species of plants capable of converting solar energy, carbon dioxide from the air, and water into complex organic substances that form the basis of the food chain. Some 10,000 of these fit somewhere into the scheme of human economic activity, including about 3,000 that have been grown for food. Only a few of these, however, are of major commercial importance— one hundred species or so—and only about fifteen fill the bulk of the food requirements of our species. It is apparent that much remains to be discovered about many thousands of kinds of plants that could possibly be utilized.

Of course, not all agricultural scientists are engaged in research, but those who are have no lack of problems to solve. The questions in which they are interested run through the whole spectrum of biological disciplines. Sample topics for investigation might include: What do viruses do inside plants? What causes changes in fish population? What influences learning and behavior in animals? What is the importance of bacteria in the stomachs of cows, deer, and other ruminants? What causes plants to flower and set seed? What happens when seeds become dormant, and how can dormancy be broken most readily so that seeds can germinate better? What nutritional factors in animal feed influence the quality of meat? How can depleted soils be best brought into production? What factors in air pollution affect the growth and quality of plants? The questions are endless.

Individuals trained in the sciences related to agriculture, horticulture, and aquaculture have many career opportunities in addition to research. For example, one of the nation's largest industries is the production of livestock for meat. There are managerial and production jobs in this industry that benefit from the knowledge of physiology, nutrition, genetics, and range management that the bioscientist provides. Skills in engineering and business management will add to the value of the person seeking such jobs. Animal and plant products also undergo further processing before being offered to the consumer, and many thousands of trained people are employed by processing plants. Careers also exist in the communications aspect of agribusiness. Such careers require, in addition to technical competence, facilities in writing or public speaking. A thorough knowledge of English is a valuable asset for any career, but it is especially necessary for the many jobs available in trade publications, livestock magazines, advertising agencies, and many other agencies servicing the industry. Individuals trained in animal science advise food processors, act as consultants to engineers who design machinery and equipment used in agriculture, assist breeders of horses, are involved in breeding and maintenance of laboratory animals, and help in the production and testing of chemical products for use in animal or human nutrition or medicine—to mention just a few applications of their skills.

The land-grant colleges located in each state, as well as many other colleges, including some community colleges, offer training in animal science or other science applicable within the context of the above discussion. A typical course includes, in addition to the basic foundation in humanities and biological and physical sciences, applied subjects such as animal breeding, reproduction, nutrition, and various aspects of managing and handling animals. Many programs also offer a number of support courses in such areas as food science, forage production and use, crop production, and soil science. In addition, communication skills and the economics of agribusiness may be covered, including computer science. The learning is not all from books, for most such schools maintain flocks of poultry, herds of livestock, dairies, and even facilities for practicing the production and packaging of foodstuffs and other products. Practice also is afforded in the all-important paperwork. These specialty courses, of course, come after one has obtained the sort of background that any well-educated person should have in basic science, mathematics, languages, and the humanities.

After obtaining such training, employment opportunities may be sought in many areas, including some of the industries mentioned above. Private research foundations employ many scientists with training in applied biology, as do industries engaged in service to agriculture, food processing, and forestry products. The agricultural scientist may work in the classroom, laboratory, or experiment station. Agricultural colleges employ many teachers and research workers in biology. In order to fill such jobs, the student will generally need education beyond the baccalaureate degree, and many of the colleges offer graduate work leading to advanced degrees. Research agencies of the federal government and state agricultural experiment stations employ many thousands of agricultural scientists. Those in charge of the various projects typically have doctoral degrees, but there are many rewarding jobs that can be filled by those with baccalaureate or master's degrees. The person wishing eventually to obtain a doctorate often can fill the junior positions on the staff while working for a more advanced degree.

FORESTRY

Like agricultural science, forestry is a highly practical subject. Indeed, in some lists, forestry is classified with the agricultural sciences. A forester must have training in both the scientific and practical aspects of the management of forest lands. Timber is an important biological resource, and it is the forester's job to see to it that it is used and yet perpetually maintained. The importance of forest products in our economy may be emphasized by citing a few figures: forest lands cover approximately one third of the total land area in the United States. Much of this is set aside as United States National Forest, in which tree harvests are regulated by the federal government. There are National Forests in thirty-nine states and Puerto Rico, and their total area amounts to about 15.6 percent of the area of the United States. To get an idea of how large this is, if the national forests were all in one continuous plot, it would cover all of New England, plus the states of New York, New Jersey, Pennsylvania, Delaware, Maryland, Virginia, West Virginia, Ohio, and Kentucky—and still have enough left over to cover two-thirds of the state of Indiana. In addition to the national forests, many states have state forest areas, and all the big timber companies have additional huge tracts of their own.

Forests are a rich source of products, the most obvious of which is wood; but forests also supply various berries and nuts and the material necessary for making maple syrup, paper, cardboard, some pharmaceuticals and plastics, and many other goods. The forest products industry in the United States makes about $48 billion worth of products annually, and it employs about 1.3 million people.

Foresters, like plant pathologists, are concerned about plant diseases. Like entomologists, they must be familiar with insects and their effects, and like ecologists, they must understand the whole of the forest environment and be interested in conservation and the management of wildlife. Foresters must have some expertise in all those fields. In addition, some foresters must learn how to manage outdoor recreation areas. An understanding of many of the technical aspects of wood products, pulp, and paper is also important.

Much of the forester's work is out-of-doors, but there is a great deal more to it than a perpetual outdoor hiking trip—much of the work is arduous and hazardous. As the forester gains experience, there may be more administrative and less fieldwork to do.

A bachelor's degree in forestry is the minimum educational requirement for a career in forestry, and many employers prefer graduates with advanced degrees. Curriculas stress liberal arts, communication skills, and computer science in addition to technical forestry subjects. Many colleges require students to complete a field session in a camp operated by the college. There are about fifty-five colleges in the United States offering courses leading to a bachelor's or higher degree in forestry. Most such programs are accredited by the Society of American Foresters.

To give some idea of the variety of jobs available to one trained in forestry, consider this partial list: lumber inspector, forest ecologist, forest products technologist, wood products salesperson, forestry information specialist, campground supervisor, forest pathologist, urban forester, Christmas tree grower, wood fuel expert, tree geneticist, forest pest controller, forest consultant, and tree service expert.

Although the federal government is the largest employer of foresters, many states, cities, universities, and private corporations also offer employment opportunities. Private employers include lumber, pulp, paper, and other manufacturers of wood products; also railroads, electric utilities, water companies, recreation clubs, and owners of large private estates. Several departments and bureaus of the federal government employ foresters, but the majority of such employees are in the U.S. Forest Service.

LABORATORY ASSISTANTS, TECHNICIANS, AND TECHNOLOGISTS

In any work that a scientist does, whether in a service laboratory, a research institute, a university or hospital, or in industry, he or she is part of a team. The leader of the team may be a highly trained scientist with a doctoral degree, but the work can progress only with the skilled and interested cooperation of technically trained coworkers. Technicians are the "doers," performing the practical operations, leaving theory to the scientist. Sometimes the technician is in training to become a scientist, but often the job of technician is regarded as a fully rewarding career in its own right.

The technician operates the laboratory equipment, often designs and constructs new equipment, makes drawings, builds models, estimates costs, makes repairs, keeps records and assists in the interpretation of results. The positions offered are found by many to be pleasant and profitable, and there is often an opportunity to participate in research, if the technician is so inclined. So great is the variety of the work performed by technicians and technologists that the National Health Council has estimated that more than 200 such careers are available in the health field alone.

What makes a good technician? Let us attempt a partial list:

• Some background in science and mathematics, focused primarily in the practical direction of problem solving.

- Ability to work with the hands.
- A practical bent.
- Work habits that are systematic, precise, and patient.
- Ability to work under pressure.

Health service workers assist fully qualified health professionals such as physicians, dentists, and veterinarians in the handling of patients. For example, the physician assistant may perform, under the close supervision of a physician, many tasks ordinarily performed by the physician, including obtaining case histories, scheduling laboratory tests, handling emergencies, and other clinical procedures—indeed, the physician assistant works so closely with the physician that he or she is known as a paraprofessional and is officially recognized by the American Medical Association. The qualified health workers of which the physician assistant is an example constitute a large, growing, and important group of careers.

Technicians and technologists require training beyond high school, but much of the skill required in these jobs can be obtained only on the job. There is a wide range of requirements for formal schooling. Medical technologists perform complicated bacteriological, chemical, microscopic, and other tasks; they usually have at least four years of college and a baccalaureate degree. A similar training, though less complete, with two years of postsecondary schooling, may qualify one to be a medical laboratory technician. The technician displays a high degree of skill and performs a wide range of tests but has less technical and theoretical knowledge than the technologist.

Sometimes laboratory technicians have the opportunity to study for advanced degrees and become professional bioscientists of the highest rank. Not all technicians entertain ambitions for such advancement, and many find satisfaction in a job well done in support of the laboratory's work. Individuals who fill these positions must have some training in theoretical and applied biology; they are recognized professionals or paraprofessionals, and they serve important functions.

Large numbers of workers are employed in medical laboratories supervised by pathologists. They receive special training that entitles them to be members of the Registry of Medical Technologists. Several levels of laboratory workers, depending upon the amount of training necessary, are recognized by the American Society of Clinical Pathologists (ASCP) and the American Society of Medical Technologists. The certified laboratory assistant (CLA) receives accreditation after one year's study beyond high school in a laboratory assistant school accredited by the American Medical Association. The histologic technician is a high school graduate who has spent one year after graduation in a hospital training program or junior college. The job consists of preparing microscope slides for study by the pathologist. The CLA and the histologic technician may obtain employment in a clinic, physician's office, public health agency, or a variety of industrial, pharmaceutical, and military laboratories. The cytotechnologist works in a restricted specialty involving microscopic study of cells to assist in the diagnosis of cancer. This specialty requires two years of college, an additional year in an accredited school, and practical experience. Scholarships are available for students in cytotechnology.

Medical laboratory technicians generally have an associate degree from a community or junior college or a diploma or certificate from a trade or technical school. Some people also become medical laboratory technicians through on-the-job experience, specialized training, or a combination of these. Medical laboratory technicians perform a wide variety of routine tests under the supervision of a medical technologist or other laboratory supervisor.

Medical technologists perform more advanced duties than do medical laboratory technicians, and they need at least a bachelor's degree in medical technology or one of the sciences. Programs in medical technology include courses in chemistry, general biology, microbiology, and mathematics, as well as course work devoted to acquiring the knowledge and skills used in the clinical laboratory. Some programs require the student to take courses in management, business, and computer science. Medical technologists are in great demand, and their work is varied and interesting, involving the performance of chemical, microscopic, bacteriological, and other tests. The practitioners of this specialty work in laboratories connected with physicians' offices, hospitals and clinics, independent service laboratories, government agencies, and industry. They often advance to positions as supervisors, teachers, or research assistants.

Many of the schools where training for these jobs can be obtained are in state-supported universities, but there are also many private technical schools that specialize in this field. An up-to-date list of the accredited private schools is contained in the *Handbook of Trade and Technical Careers and Training,* which may be obtained free of charge from the

National Association of Trade and Technical Schools, 2251 Wisconsin Avenue, NW, Washington, DC 20007. Most of the members of this association offer training totally unrelated to bioscience, but many do include complete courses in laboratory technology.

There are many categories of workers in medical laboratories in addition to the above. Indeed, it has been said that of all the fields hiring technicians, medicine offers the greatest diversity of careers. To mention just a few more: an EKG technician learns to record electrocardiograms for the physician after a training period that is usually a few months to a year. A somewhat longer period of training—up to two years—is undertaken by the EEG technician, who learns to record the electrical activity of the brain through the use of an instrument called an electroencephalograph. Instruction in anatomy, neurology, medicine, psychiatry, and electronic instrumentation is supplemented by laboratory experience. Radiologic technologists not only take and develop x-ray pictures, but help treat patients by administering radiation therapy; they work closely with radiologists, and the course of training takes from one to four years. The microbiology technologist must have a bachelor's degree, preferably with a microbiology major, followed by a year's experience beyond college. The blood bank technologist also requires a baccalaureate degree, preferably with a major in biology, followed by a year in a school approved by the American Association of Blood Banks.

There are many areas of biological sciences that employ technicians besides those related to medicine. Practically all scientists need technical assistance. In some instances, a high school graduate who has a good record in bioscience and other sciences and a strong desire to learn on the job may obtain one of these positions directly, but more often the employer would give preference to one with further education. There are many opportunities for technicians with a baccalaureate degree with a biology major, even without any specific training as technician; it depends upon the flexibility of the employer. A number of technical institutes offer training to qualify students for a job immediately after graduation. These curricula, some of them in junior and community colleges, are similar to the first two years of a four-year college, and the graduate may either qualify for a technical job then or transfer for further education.

There are specific courses for training food science technicians, marine science technicians, natural resources conservationists, animal husbandry technicians, plant science technicians, and soil technicians.

As an example of the kind of work one of these technical assistants might do, a plant science technician, after one or two years training, might assist an agronomist (one skilled in the scientific management of crops) in experiments on basic nutrition of plants, on breeding and selection of plants, or on a study of plant diseases or relationships with insects, or in discovering new economic uses of plants. Chemical analyses of plants and the collection and preservation of specimens might be other objects of attention. Other biological scientists—ecologists, zoologists, physiologists, pharmacologists, biochemists, biophysicists, and many others—also benefit from the services of technicians.

Animal technicians assist veterinarians and research workers in biology. They care for and feed animals, keep records, maintain equipment, prepare animal patients for surgery, collect specimens for laboratory analysis, dress wounds, and perform many kinds of laboratory procedures. In recent years, there has been increasing recognition of the need to upgrade the technical skills of the workers who handle animals in biological research laboratories, veterinary practice and research, or any area where science and animal care are combined. Especially active in promoting improvements in animal care and handling are the American Veterinary Medical Association (AVMA) and the American Association for Laboratory Animal Science (AALAS). According to a publication of the AALAS, a laboratory animal technician is a person in a position requiring education below the doctoral level.

The AALAS seeks to improve the abilities and status of animal laboratory workers by developing standards of training and by recognizing individual accomplishment through a three-level program of certification. The animal caretaker, with the title of assistant laboratory animal technician, provides the day-to-day maintenance of the animals and their quarters under supervision. After completing a course (which may be on the job) and then one year of experience, the caretaker is eligible to take an AALAS-sponsored examination for certification. The next grade, laboratory animal technician, requires three years of experience, two of which may be spent in a laboratory animal technology school after graduation from high school. The laboratory animal technologist may have a baccalaureate or master's degree in animal science or a related technical program. Work experience of six years may be allowed in lieu of a degree; in any event, written examinations must be passed.

WRITING IN THE BIOLOGICAL SCIENCES

No matter what field of bioscience you are in, or what the level of your career, you'll find times when it is to your advantage to be able to write well. The technician must keep clear and accurate notes of his or her observations and may be called upon to submit reports to the scientist in charge of the laboratory or the field trip. The professional scientist will discover that as one's career advances, the demands for writing become more and more insistent. Some may wish to communicate with the public as well as with fellow scientists and may wish to make a career of writing. Articles on science are popular with editors of newspapers and magazines, but few individual papers employ science writers; most of them buy articles on science from syndicates.

In order to advance scientific knowledge, it is essential that new discoveries be reported in such clear language that the discoverer's colleagues can understand it—and if necessary, repeat the observations. Indeed, a scientist's reputation is largely based on such communications. For this reason, the biologist who has writing ability has an advantage over one who does not. Most scientists—as well as most other people—find writing to be hard work, and no one ever becomes able to communicate *all* of one's thoughts to another person. Nevertheless, there are some who become so skillful at it that it becomes a pleasure (even though it remains hard work), and some devote all their time to it.

Practically all scientists do some writing as part of their career work. Universities require candidates for doctoral degrees to write dissertations; these are often very long and elaborate descriptions of the work done, the reasons for undertaking the job, the historical background, and the interpretation of the results. This is, in a way, practice for what the scientist will do in the future—the research worker, especially in a university, must write papers reporting the results of the research. Unless this is done, the research has not really contributed anything to knowledge, nor is the worker likely to receive rapid promotion in the job, no matter how skillful the worker is in other respects. The worker outdoors in the field often makes observations that need to be communicated to colleagues. The educator may write reports on new teaching techniques or on especially informative experiences in the classroom. If you intend to be a professional scientist, you should start learning to write well as soon as possible.

The scientist who has written a number of such reports will sometimes undertake a more ambitious writing project in the form of an advanced treatise or monograph. A monograph is a highly specialized book attempting to cover what is known about a specific—and usually rather narrowly circumscribed—aspect of a subject. A monograph may have a single author, or different authors many write the various chapters in the book, with one of them serving as editor of the entire monograph. If a scientist has acquired a reputation for writing research papers, it is likely that at some point in her or his career the person will be invited to write some sort of review or other monograph.

Biologists also write laboratory handbooks and textbooks intended for class use. Most publishers send each of their new scientific books to the editors of *American Scientist* and other professional journals in the hope that they will receive favorable reviews. The number of such new books received for review each year exceeds a thousand, and the editors of *American Scientist* engage more than 350 scientists to read and review the books received.

In a somewhat different category is science writing done for the general reader. This may take the form of articles for newspapers or magazines, columns reporting the latest developments, interpreting science news for the public, or answering reader's questions. Professional science writers particularly skilled in summarizing current trends in scientific discovery and thought may find employment on the staff of a magazine devoted to popular accounts of science—such periodicals as *Science News, Popular Science,* and *Science Digest,* for example. A different kind of science writing is to be found in *Scientific American,* which publishes semitechnical articles, often by scientists in one specialty explaining their work to those in unrelated fields. Popular nontechnical books addressed to the general reader also are much in demand.

In the past, scientists in general and biologists in particular have been critical of much of the reporting of science news in newspapers and magazines, feeling that many of the reporters thought only in terms of the human interest of their stories instead of scientific accuracy. This handicap, coupled with the lack of cooperation from biologists who often held themselves to be above talking to representatives of the press, sometimes led to misleading stories. Mistakes still occur in some science reporting, but in general there has been a trend for the press to hire science reporters who have had training in biology—some of them even hold advanced degrees. Such people are truly professional science writers, and they have an organization whose aim is to maintain high standards. There also has been a change in the attitude of many biologists toward the press; many meetings of biologists now feature newsrooms where scientists and reporters can meet together, and some societies prepare news releases describing in nontechnical language some of the reports presented by the scientists to their assembled colleagues.

Generally speaking, the biologist who writes reports of work for the information of colleagues for publication in scientific journals receives no monetary compensation for these efforts. Technical journals appeal only to a limited readership and cannot afford to pay their authors; indeed, often the author or the scientist's institution is required to pay *page charges* to the journal. The author of an advanced technical manual or monograph fares somewhat better—but not much. These highly specialized books cost just as much to produce as do basic textbooks, and sometimes more. But it is easy to see that a monograph with a sale of 4,000 copies will cost much more per copy than a textbook that sells 10,000 copies. Not only does the monograph have a limited appeal, but its high price further reduces sales. Hence, even if the author has a favorable royalty agreement with the publisher, the scientist would not consider writing a technical book about her or his specialty as a means of income. The rewards must be the more intangible ones of prestige and the personal satisfactions that come from a job well done and from seeing one's work in print.

Writing books for the general public, on the other hand, or writing textbooks for class use, can be profitable either as a sideline or in some instances as a full-time career. Beginning authors sometimes have illusions about their writing abilities, the importance of their manuscripts, and the financial returns to be expected. Skill in writing sufficient to turn out acceptable manuscripts requires years of practice beyond what can be learned in a freshman composition course. Although publishers will read unsolicited manuscripts, the most favorable reception will likely be given to manuscripts that the editor is expecting. The groundwork may be laid by previous correspondence between editor and author, but authors who are interested in writing as a full-time career usually employ literary agents.

CHAPTER 7

PREPARING FOR YOUR CAREER IN THE BIOLOGICAL SCIENCES

TRYING A CAREER IN BIOLOGY ON FOR SIZE

Now that you have read about the many varieties of careers in the biological sciences, perhaps you have an idea of which career is for you. The education of a biologist requires a large investment of time and money. So before you commit yourself, why not "take a test ride?" Fortunately there are several ways to test your perception of what your job choice would be like against the reality of how you feel when you actually are working in that environment with people who could be your coworkers.

In the first section of this chapter, we will look at volunteer or part-time work and internships as ways of trying a career out by actually doing it. In the next section, we will look at another approach, using a personal computer to log onto Internet World Wide Websites to communicate with real scientists and scientific organizations. "Surfing the web" is a way of experiencing your job in "virtual reality." A combination of real and virtual time spent visualizing yourself as a biologist is a smart way of trying on a career before finding out that it's not for you in the senior year of an undergraduate program.

Volunteerism is in! Many people are finding that giving to other people and organizations enriches their lives in ways that paychecks can't. First think about where you would like to work. Without a Ph.D. you may not be able to do the job you would like to have someday, but if you can get your foot in the door as a volunteer, you would have a chance to mingle with scientists who are what you want to become. Also, if you have never worked in this type of environment, it would give you some experience in acclimating to the work schedule and other aspects of the organizational culture. You may have to start at the bottom of the career ladder—type and file paperwork, clean cages, answer the phone, give tours to visitors, straighten the lab, and so forth. But it's a chance to keep your eyes and ears open and to ask questions to your heart's content! You can find the places you want to work in a number of ways. Look in the Yellow Pages of the phone book; contact some of the associations listed in the appendixes and ask for a list of members or sites that might have volunteer opportunities; go to your local public library and ask the librarian to help you find a list of volunteer organizations or a Chamber of Commerce directory in your area that would list the type of organization you are interested in; ask you biology teacher and guidance counselor.

Part-time jobs are becoming more plentiful and better paid. The first place to look might be the employment section of the newspaper or the jobs bulletin board at school. You also might try the same sources listed above for volunteer opportunities, but call to ask for part-time work instead. One of the best strategies for job hunting, part-time or full-time, is "networking." Tell everyone you know that you want to find a part-time job that will help you try your intended career in biology on for size! You might be surprised to know someone who knows a biologist who needs a part-time helper! Prepare a short one-page resume that outlines your qualifications for the job. Include such things as volunteer work, sports, committee work, offices held, and courses taken at school. Take a job in the company or organization that employs the type of professional you wish to become and then find that someone. Ask that person to be your "mentor," or someone who will give you the personal and professional advice to become the type of biologist you want to become.

Internships are generally educational experiences that are designed to give the participant exposure to a future career. Summer internships are becoming more popular with nonprofit organizations that need help, but have limited funding. For you, an internship is an excellent opportunity to try a career, because instead of pay, you can structure an educational experience that might expose you to more experiences than a part-time or volunteer job. Many organizations use interns to help with special projects that they do not have staff or time to accomplish. As part of the educational experience, you will probably be required to write a paper, do a presentation, or complete an assigned project. Your high school or college also may grant academic credit for your internship, if it is approved ahead of time. You can locate internships by contacting the professional organizations listed in the appendixes of this book or by calling an organization where you would like to work. If the organization does not have an internship program, ask to meet with a staff member who needs help with a special project and submit a proposal for an educational experience that would help both you and your mentor. Who could resist a "go-getter" intern like that?

Don't forget to keep a record of exactly what you do as a volunteer or part-timer. This experience will form the basis of a resume that you can use for later jobs. It will show your early commitment to your chosen career. Make sure that you keep up-to-date information about people you work with in the jobs who can provide references.

THE VIRTUAL BIOLOGIST: SURFING THE NET

In the appendixes and throughout this book, there are references to Internet World Wide Websites (www). If you are not computer literate or have never used the Internet, it is time to explore this new and exiting resource. You might wonder why some of your friends talk about getting "hooked on the net." It is because it is so interesting and so much fun that you won't want to stop. And it's easier than you think to use it. You can get the training you need at school or check with your local library. Larger public libraries frequently have computers where you can learn and use the Net. Much of the new computer software that is designed for accessing and searching the Internet is "intuitive," or easy to use, with occasional reference to built-in "help" information.

Once you know how to access the Web, you will find that typing "biology careers" into your search software will connect you with some of the most interesting sites that contain pages of information, including color photographs. This information can be downloaded into your personal computer and then printed. One of the marvels of the "information highway" or Internet is it's unique ability to be interactive in real time. It's the next best thing to the telephone! You can log onto a site and send a message via E-mail or to a live chat room. Recently, the news was full of reports of schoolchildren sending E-mail via the Internet to astronauts who were performing science experiments in space. Can you imagine how exciting it must be to experience the excitement of space exploration in real time, millions of miles away, instead of reading about the results of the experiment in tomorrow's newspaper? Back on earth, you can interact with researchers on the cutting edge of the biological sciences or other discipline by logging onto a website like the Scientific American. You can ask questions or read their responses to others. It's like being there!

Other ways of experiencing the world of the biologist before you take the plunge is to read what they read. You can use the Internet to peruse the full text of many scientific periodicals and documents without traveling to a university library. Almost every college and university has its biology department's curricula and information about student life available on its web page. You also can review the credentials and interests of faculty members who might become your role models. Sometimes it is possible to access the students' homework assignments or review their reading materials to get an idea of the scope and difficulty of a particular course.

Posting your resume online is a popular way to job hunt these days. You can compare your experience and plans with those of young biologists who are upwardly mobile.

Since Internet addresses change a lot, some of the resources we have suggested here may not be accessible by the time you get to try them, and as soon as this book is in print, there will be a host of new websites. It's hard to keep up, so our best advice is to learn to use a web browser or search engine so that you can select from everything that is available when you need it. A list of favorite websites for general information on biology and biological sciences careers is included in the appendixes. Explore these and make your own list of favorites.

NAVIGATING YOUR EDUCATIONAL ROAD MAP

According to the *Occupational Outlook Handbook,* published by the Bureau of Labor Statistics, the outlook for employment opportunities in the biological sciences will remain good through 2005, especially for those with advanced degrees. If you are considering a career in bioscience, you will find it to be to your advantage to obtain as much education as possible. Most of the careers to be discussed in this book require education beyond the high school level. Graduate study beyond the baccalaureate will be advantageous and indeed will be necessary for the topmost jobs.

It will be seen in succeeding chapters that there are many biologically oriented careers other than research or teaching for which advanced degrees are not essential. Professional biologists who wish to have the greatest choice of jobs may wish to climb the traditional ladder that includes a high school course with several science subjects, then a college course with broad training in the first years followed by a concentration on the major subject, then after graduation a more intense concentration on a narrower aspect of the major through the acquisition of a master's degree and finally a doctorate. All these steps are important for those who will fill the highest positions as research scientists or university professors. Indeed, some have found it advantageous to go even further, through a period of postdoctoral training—somewhat comparable to the internship and residency requirements of a medical specialist.

There are a host of problems facing our society that can best be solved by those with some biological orientation, and one can scarcely be considered to have a well-rounded education without some knowledge of bioscience. There are also a large number of biologically oriented jobs in certain industries—some manufacturers of food products, for example—that do not require the specific training of a biology major.

If you are considering a career in biology, the level of education you pursue beyond high school depends upon what place you wish eventually to assume on the biological team. If you wish to work in a laboratory, there are schools that take high school graduates and train them in a couple of years to be laboratory technicians. Their graduates have no trouble obtaining jobs with good pay. They work in doctors' offices, hospital laboratories, or various research centers. For advancement, however, further education must be obtained. For one of the best jobs as a laboratory technician, you will need to finish college.

It is not unusual for a young high school or college graduate to accept a job as laboratory assistant or technician with the intention of continuing study as time permits, earning credits toward a degree. Some have gone all the way through college and on to the Ph.D. degree this way. It may be done by part-time schooling or by alternately working and studying full-time. Some employers, especially industrial laboratories and some government departments, encourage employees to improve their education by permitting time off for class attendance and by paying tuition bills.

The baccalaureate degree qualifies one for many teaching jobs in elementary or secondary school, but the additional one or two years of study needed to obtain a master's degree offers an additional advantage. At least an additional two or three years are required for the doctorate. With exceptional good fortune, the doctorate can sometimes be earned within three years after graduating from college, but most graduate students in bioscience receive the Ph.D. degree seven years after the baccalaureate. In the meantime, they have had three years of professional experience. Individuals with the doctoral degree are equipped to fill the most exciting and challenging jobs in research. The best jobs in industry, government laboratories, independent research institutions, or university faculties also are available to those who have pursued their education through the doctoral level.

This may seem like a long and discouraging prospect, but remember that during all those years you will be learning things that are beyond anything you have dreamed of. If need be, you can be earning while learning. In the end, your rewards will be not merely an adequate livelihood, but satisfaction in a life's work that no money can buy.

HIGH SCHOOL SUBJECTS

It has been said that the scientist speaks with three kinds of languages: English, mathematics, and foreign languages. High school is not too soon to start the mastery of all three. English is the language you will be using to communicate with your fellow students and with your fellow scientists; you should strive to speak and write it correctly, concisely, and with the precise meaning you wish to convey. Unfortunately, many scientists are deficient in this regard. Editors of scientific journals frequently complain about the quality of the writing in papers submitted to them for publication. If you can learn to speak and write clearly, you will find your path smoother no matter in which direction you wish to go. Your colleagues will appreciate and respect your ability to communicate. So study as much English as possible, work hard on composition, and read, read, read.

The use of mathematics varies greatly among the various fields of biology, but more and more biologists need to know calculus and statistics, at least. For some of the newer fields of research, such as molecular biology and biophysics, more advanced mathematics is required; the mathematics requirement is more fully discussed in the next section of this chapter.

Science knows no international boundaries, and much of importance to biologists—especially if they are engaged in research—is published in languages other than English. At international gatherings of bioscientists, it is necessary to communicate with people who speak other languages. For these reasons, certain language requirements have been established for students majoring in biology or other sciences. The high school student intending to go on through graduate school should anticipate these requirements and start language study as early as possible. If language study is postponed, it would require time that advanced students would prefer to devote to their specialty.

In addition to English, mathematics, and language, high school students interested in any science should include chemistry, physics, computer science, and biology in their programs. Most high schools offer computer science. Good skills in the use of the personal computer will be helpful in college and your career. Many computer users find a typing course to be an invaluable prerequisite to efficient computing. With these studies, together with the basic requirements for graduation, the student will enter college well prepared to pursue any course of study— biology, another science, or nonscience. If after discussing your plans with your guidance counselor you feel unsure about whether you wish to prepare for a career in biology, or for that matter, if you don't even know what your major in college will be, you should not feel discouraged. It is too frequently not recognized that is the exception rather than the rule for students at the beginning of college to be certain what their future will be, and many young people discover new interests after they have been in college for one or two years and change their career goals then or even later. If you have a proper foundation of knowledge in English, mathematics, languages, science, and the humanities, you are prepared to undertake a study in any field of endeavor.

COLLEGE REQUIREMENTS

The choice of a college presents no unusual difficulties for the young person intending to study biology. All—or nearly all—of the better colleges have departments of biology that range from adequate to excellent.

In most instances, the choice of a college for undergraduate study may safely be made on grounds other than the size or research reputation of the biology department. Of course, the biology department of the chosen college should be able to present a good solid major in biology; one way to judge its adequacy is to inquire whether the graduates of the department have been accepted in programs for advanced degrees at outstanding universities.

I cannot emphasize too strongly that you should not feel compelled to consider only Ivy League colleges or others of similar age or social standing. They are fine schools, and their reputation for excellence has been well earned, but they do not have the monopoly on academic achievement that some status-conscious but ill-informed people may believe. Many of the outstanding biologists of recent years were graduated from colleges that are little known to the general public outside of their nearby communities. These colleges are widely dispersed geographically, and there is no one section of the country that can claim an outstanding advantage. Indeed, several surveys have been made of the academic origins of scientists. The surprising finding is that the small colleges of liberal arts that are scattered in such profusion throughout the country contribute a disproportionately large number of biologists and other scientists.

In your preparation for a career in biology, you will continue in college many of the subjects that you studied in high school. You will be able to plan your course of study according to your individual needs and preferences; you may have a member of the faculty assigned as your specific adviser. In this chapter, we can give you no more than a few general statements about the kinds of studies you will need. During the first two years, most students will take courses in a wide variety of subjects to obtain a broad education. In the last two years, they specialize more and more in the major field, with a somewhat less thorough specialization in another (usually related) field called the minor. Most biology majors have minored in chemistry, mathematics, or physics, though in recent years some have minored in anthropology, sociology, or psychology.

The specific courses in bioscience that a biology major takes will vary depending upon which branch of biology is chosen and upon those offered by the college. Generally speaking, the larger colleges offer a greater variety of courses. In some instances, separate majors may be offered in various fields of biology, such as botany, microbiology, or zoology. For most students, it would be unwise to specialize too strongly at an early stage. A career brochure once prepared by the AIBS put it this way: "Give yourself a chance to understand the breadth of biology before you concentrate on the blood flow in the muscles of a frog's leg." Some kind of a *core* course is often offered to give the student a bird's-eye view; you should familiarize yourself not only with the different branches of bioscience but with the careers offered in applied biology as well.

After deciding to major in biology, most students will seek to know something about cellular biology, study the fascinating facts of heredity (genetics), learn how the organs of the body function (physiology), obtain some knowledge of systematics and ecology, and find out what is known about the changes that occur with the passage of many generations (evolution). Comparative anatomy, embryology, and microbiology also will be on the list. Even if the student does not take a specific course in each one of these subjects, it is essential to become aware of the general principles of each of them, regardless of what biological speciality is eventually settled upon. After finishing, the student will find that he or she is equipped not only to go on to graduate school but to enter the study of such specialties as medicine, dentistry, veterinary medicine, agricultural science, and many other professions as well.

While the student is studying biology, the other sciences will not have been neglected. Indeed, an understanding of biology requires knowledge of other sciences. Although a wide range of options may be available to the student in the third and fourth years of college, there is a growing tendency to require a core of mathematics and physical sciences common to all options. An example of this is a statement prepared by a Committee on Undergraduate Curricula of a major university that stated that the bachelor's degree in biology "should provide an introduction to the areas of mathematics and physical sciences of sufficient intensity to permit the student to perceive and think in terms of the physical and quantitative components of biological phenomena."

What these words mean in terms of practical requirements for the student may be summarized somewhat as follows: After a preparation in high school, including introductory chemistry and at least one and a half years of mathematics, the student continues mathematics through calculus, analytical geometry, and elementary statistics. For some biological specialties, that would not be enough mathematics—some biophysicists recommend as many as ten courses in mathematics for their specialty.

General chemistry, organic chemistry, and physics must also be a part of the curriculum for those intending to become professional biologists at the graduate level. Furthermore, you should learn how to use computers.

These requirements will vary somewhat with different colleges, but by and large you may be expected to be prepared to fulfill most or all of them. It was not always so. Some very successful older biologists in certain fields—notably anatomy, botany, and zoology—received their training when such a rigorous schedule was not in vogue. Yet, even in these specialties you will find that the actively working bioscientists will recommend that their students become fully equipped with the tools provided by mathematics and the physical sciences.

The above does not complete the list of subjects that biologists may need. Some of these will be mentioned in succeeding chapters when some kinds of biological careers will be discussed. Biochemistry, psychology, and geology are among the subjects that will appeal to some biologists. Students intending to study for advanced degrees also will expect to be required to show some facility in reading biological literature in one or more foreign languages. Language study should start in high school and continue in college. Language requirements have changed considerably over the past generation or so. Not long ago, all high school students planning to go to college took two or more years of Latin in addition to a modern language. Nowadays, no college requires Latin for admission, and most high schools do not teach it. Probably most of the Ph.D.s in bioscience obtained their degrees when reading knowledge of both French and German were required. Gradually, it become possible to substitute another language—Russian or Spanish, for example—for one of the traditional languages in special fields. Even that has been changed in many schools, with only one language required. In certain fields (fisheries research, for example), more current research is published in Russian or Japanese than in French or German.

The language requirement for the doctoral degree can be met with less knowledge of a language than is necessary for fluency in speaking, reading, or writing, but the student would find it advantageous to acquire fluency, if possible. There are increasing numbers of international meetings attended by biologists from many countries. Although most of these people speak English (with varying degrees of proficiency) and the majority of the talks at many meetings are in the English language (all educated continental Europeans seem to be better linguists than the average educated American or Briton), anyone attending finds it most profitable to possess more fluency in a foreign language than the minimum necessary to pass the usual college examinations.

I would not want to close this discussion of the undergraduate education of a biologist by leaving the impression that the student need study only English composition, sciences, and languages. A biologist is a member of the intellectual community and cannot properly fulfill her or his function in society—or even be a good biologist—without being an educated person. The student biologist should, therefore, take advantage of the opportunity to learn as much as possible about history, philosophy, economics, religion, the arts, or whatever liberal arts subjects have most appeal or that are recommended by the adviser. In today's world, the truly educated person must have more than a little exposure to *both* science *and* the humanities.

GRADUATE STUDY

In many foreign countries, there is no exact equivalent to the baccalaureate or master's degrees, but the doctor of philosophy is universally recognized and respected as the highest official reward for the completion of a course of study aimed at the training of scholars, teachers, and research workers. In the United States, the first Ph.D. program was initiated by Yale University in 1860. There are some variations in the requirements from one university to another, but essentially they all include completion of certain prescribed courses; fulfilling requirements for a reading knowledge of one or more foreign languages of material in the candidate's major (this requirement has varied more in recent years than have the others); writing a dissertation (thesis) that presents and discusses the results of the student's original research; and passing special examinations, usually consisting of written examinations testing the student's broad knowledge of the subject and a final oral examination centering upon the points covered by the thesis.

These sound like formidable requirements—and they are. But students in the biosciences have been increasingly successful in satisfying them. Since the mid-1950s, when about one thousand doctoral degrees were rewarded annually in the biosciences, the number has approximately tripled. The young people obtaining these degrees are equipped, insofar as formal education can equip them, to obtain desirable employment at once and ultimately rise in their chosen profession as far as their abilities and achievements will carry them.

The degree of doctor of philosophy (Ph.D.) has a venerable history and may be rewarded in many fields having no relation to science. Some universities, wishing to emphasize the concentration on science, award the degree of doctor of science (D.Sc.) rather than the Ph.D., just as in some instances the baccalaureate degree is B.S. instead of A.B. In each case, the degrees are equivalent.

In most institutions of higher learning, promotion and salary increases are based upon research success and especially upon scholarly publication. Often, teaching ability is scarcely considered, and tenure may be denied to excellent teachers who have been less than prolific in publication—a situation known among college faculty members as "publish or perish."

The route to the doctorate may seem like a long and difficult path to you, but be assured that it is really not as bad as it sounds and is probably no worse than you would need to follow in any other professional career. Excellence in any field is attained only by hard work and devotion. Musicians must spend long years practicing scales and exercises and learning musical theory. Writers must labor diligently over the typewriter even when the muse seems to have forsaken them, and they must work all the harder when their efforts have been rejected by themselves, editors, or the readers. Success in business, too, comes usually only after many years of hard work, sometimes accompanied by disappointing setbacks; top business executives usually have spent their lives in working long hours that make the traditional forty-hour week look like loafing by comparison. The difficulties in preparing for a career in biology are different, but they are not necessarily greater.

MEETING EDUCATIONAL COSTS

The cost of an education in biology at the undergraduate level does not differ essentially from the cost of any other college education at the same level. Those who elect to continue their education in graduate school usually will find no difficulty in achieving an earning-while-learning status.

It is difficult to generalize on the costs of a college education; the costs vary greatly among different colleges and universities, and in the past several years they seem to have increased each year. High school students should not wait until near their graduation to decide what kind of college they want to attend, to take stock of their financial resources, and to examine the possibilities available in scholarships and loans to help defray expenses. Students should make full use of the information and help available from their high school guidance counselor and from other sources.

When the high school graduate applies for college entrance, he or she will find plenty of competition, even if the applicant has a good high school record and a high SAT score. There are some colleges that accept only the highest-ranking high school graduates, and some of the prestigious colleges have many times as many applicants as they can admit. For this reason, some students get an exaggerated idea of the difficulties in being accepted in college. Some of the lesser-known colleges may start the school year with vacancies in the freshman class. Some of these can offer a highly rewarding four years of college, with good training in selected fields—some of them in biological sciences—at costs appreciably lower than those of the "big name" schools.

The student with a minimal budget may attend one of the community colleges that have sprung up all over the country. Attendance at a community college or at another college located within commuting distance of home may eliminate the expense of room and board. The AIBS has been especially helpful in promoting professional activities by bioscientists in two-year colleges, and many of these schools have developed programs in general biology comparable to (and sometimes better than) those in the lower divisions of some four-year colleges.

Every state has one or more land-grant colleges or universities; altogether there are sixty-seven of them. They originally received the designation as land-grant colleges from an act introduced into Congress by Representative Justin S. Morrill of Vermont and signed into law by President Abraham Lincoln in 1862. The first grants were for the establishment of "colleges of agriculture and the mechanic arts." These colleges owe their very existence to developments in biological science. By the middle of the nineteenth century, the advances in zoology, botany, and physiology, as well as in chemistry, led to a demand to establish schools for the study and promotion of scientific agriculture. The land-grant colleges were a direct response to this demand, and they were originally called *agricultural colleges*. They were founded as centers of research in applied biology, and they traditionally have strong departments of biological sciences.

Most of the land-grant colleges have become state universities, but many of them can still offer a college education at a cost that is often several thousand dollars a year less than that prevailing at a private college of comparable eminence. During the 1980s college tuition generally increased faster than the nation's inflation rate; college administrators blamed this situation on several factors, including the necessity to bring faculty salaries within the limits of other professions and to repair buildings that had long been neglected. The student faced with the necessity of meeting these costs will find, however, that the colleges have also increased financial aid to needy students. This often enables a student with limited financial resources to attend a more expensive school. The selection of recipients of scholarships and loans is usually based on need as well as on past achievements, so most of the funds are reserved for those who need them most. Guidance counselors generally have much information on financial aid and the qualifications for applying, but it may also be helpful for the student to do some independent investigation.

Scholarships, Fellowships, and Loans, a book by S. Norman Feingold and Marie Feingold, attempts to list all sources of funds available for student aid. The information in the book is supplemented by a bimonthly news service, and the entire volume is updated from time to time. Even so, the authors caution that their volume should be used only as a starting point, since there would be much information, especially local, that had eluded them. The kinds of aids are classified as scholarships, fellowships, grants, loans, scholarship loans (defined as scholarships that are converted to loans when the recipient fails to fulfill some requirement), work-study programs, awards, and prizes. Information also is given on any special restrictions that apply, such as residence in any particular state or locale, whether the student or parent must have some special affiliation, or whether there are citizenship requirements. Some funds, for example, are for certain ethnic groups, or for members of a fraternal organization or labor union, or for former military personnel.

Most colleges have special loan funds administered by the college, and several organizations have established foundations that supply money for needy students. Some religious denominations have considerable sums of money for lending to students of their faiths attending colleges related to the churches—and these include some excellent colleges. Your guidance counselor can help you to obtain information, but the initiative must come from you, and it is your responsibility to fill out whatever application forms may be required and to get them in on time.

Biology students are particularly fortunate in the number of part-time jobs available to them. In addition to the usual types of jobs that any college student may fill, some colleges employ students as part-time laboratory assistants. In some instances, the advanced biology student may even help out in the laboratory teaching of elementary courses. This provides valuable experience as well as a modest amount of financial aid. Part-time jobs related to biology are especially available at land-grant colleges. These colleges often operate agricultural experiment stations that provide both part-time jobs during the school year and full-time jobs in summer. The young biologist may even be lucky enough to find employment in a particular branch of biology that will turn out to be a major interest in life.

A word of caution should be given here. During a student's first year it is advisable not to accept employment that will take more than a few hours a week during the school year. Many people find the transition from high school to college a difficult one. The work is demanding, and new study habits may have to be developed. While this personal adjustment is going on, the student must remember that his or her first duty is to studies, and some limitations should be placed upon extracurricular activities, including employment. For this reason, some authorities believe that one entering freshman year in college should have the full cost of the first year in sight before the start of the year in the form of cash, expected help from family, scholarships, or loans, without needing to accept part-time employment. Such an arrangement is, no doubt, ideal, but not every beginning student can fulfill it.

Strange as it may seem, the financing of the graduate education leading to the doctoral degree is usually easier and simpler than that of the undergraduate years. Universities are eager to get talented graduate students for appointment as teaching assistants—or *teaching fellows,* as they are often called. In return for their help in teaching laboratory work in beginning biology courses, teaching fellows receive cash stipends and often free tuition for their graduate studies. In the larger universities with numerous graduate students, so much of the teaching is done by teaching fellows that these universities are often criticized for lack of contact between the undergraduate student and the eminent professors who are listed in the university catalog as teachers.

Biological research at universities is often financed by grants from governmental agencies, private foundations, or industrial firms. A large share of these funds is used for fellowships and research assitantships for students. Although some of these funds may be available for undergraduates, they are primarily for graduate study. Rather complete information on these sources is listed in the *Annual Register of Grant Support,* published by Marquis Professional Publications. If at present you feel that you need not be concerned about financing graduate study, it still may be comforting to know that it may be available when you need it, if your college record is good.

Undergraduates and those just contemplating college will be more interested in *The College Blue Book,* published by Macmillan. This is a monumental work in five volumes; the first two give descriptions and tables of data from thirty-three hundred colleges in the United States and Canada. Volume three describes all the degrees offered by college and subject, while volume four on occupational education tells where to train for technical subjects. Of special interest may be volume five, *Scholarships, Fellowships, Grants and Loans.* Similar information is available in *Barron's Profiles of American Colleges.* This is updated every year or two. Also kept up-to-date is the *Index of Majors,* published by the College Entrance Examination Board, listing all schools where any particular subject is offered as a major. Finally, summer jobs and internships and on-the-job training opportunities are listed in *Internships* from Writer's Digest Books. Your high school, college, or other local library should have these volumes available.

CHAPTER 8

FINDING THAT JOB

You are about to enter the final phase of your training, or perhaps you are about to start training. In either case, you may want to know: How will I go about getting a job when I'm ready? What may I do in the way of preparation so that I will not wind up with an advanced degree in biochemistry and then discover that the only job I can find is pumping gas? The more you know about how biologists go about obtaining jobs, the better prepared you will be for locating a job when you are ready. There are, of course, employment agencies, which are in the job-finding business for profit. In any large metropolitan area, the Yellow Pages of the telephone book have several pages of employment agencies listed. Many of them are specialized and may not have the kinds of jobs you are interested in, but some of them have special departments for technically trained personnel. You may have your name listed in an agency, then when a job is located, the agency collects some agreed-upon portion of the first year's salary, usually paid by the employer.

Whether you are enrolled with an employment agency or not, you should not depend entirely upon that approach. For one thing, you should not be bashful about writing to any prospective employer. Indeed, some employers welcome a personal unsolicited approach; there are usually more unadvertised jobs than advertised ones, and some employers may be impressed by the aggressiveness shown by an unsolicited application. This will more likely be the case in small rather than very large organizations, where you might have to deal with the personnel department rather than with the individual handling a job opening. It is important, of course, if you use the direct approach that your application be well prepared; you should obtain the help of your counselor.

COLLEGE PLACEMENT BUREAUS

Many colleges have placement bureaus that assist their students and graduates in finding jobs. These may be especially useful in finding teaching positions. Companies often send recruiters to campuses for employment interviews. Although many of these recruiters are looking for graduate students about to complete their requirements for a doctorate, openings at a lower level also are sometimes available in these interviews. Companies with vacancies in their research laboratories or in other departments will sometimes ask a likely candidate to visit them at the company's expense. The candidate may be wined and dined and given the red-carpet treatment if he or she has some special talent the company is looking for, but even a beginner can have an interview, not only with the campus recruiter, but often directly with the employer.

PERIODICALS

Many professional, scientific, and trade journals have classified sections with advertisements for "Positions Wanted" and "Positions Open." Among these are such prestigious scientific journals as *Science, Nature, BioScience, Genetics,* and *Federation Proceedings, Federation of American Societies for Experimental Biology.* Others are special trade periodicals; among those in this category that announce jobs of biological interest are *American Fruit Grower, American Vegetable Grower and Greenhouse Grower, The Commercial Fisheries News, Pulp & Paper, Wines and Vines,* and many others. Then there are those of regional interest, or those aiming their appeal to special groups, such as *California Farmer, Pacific Coast Nurseryman and Garden Supply Dealer, Southern Florist and Nurseryman, The Black Collegian Magazine,* and *Black Careers.* The above are just a few selected examples. A more complete list—in fact, some 166 periodicals overall—is contained in the *Life Science Jobs Handbook.* The list covers 23 different biological disciplines that have been the subjects of job announcements in these periodicals.

Besides these opportunities for full-time jobs, there are often opportunities for young graduates to obtain graduate fellowships; these are frequently advertised in scientific journals. Not to be overlooked are the "Help Wanted" columns of major city newpapers. Some of these, such as the *New York Times, Washington Post,* and *Chicago Tribune,* carry ads of more than local interest in their particular geographic area. Perhaps I should remind you, though, that a very popular job advertised in this way can draw many hundreds of responses, and this reduces the chances of any one applicant. It is clear, then, that in looking for a job one should not rely exclusively on such ads, but if you use them, they should be answered promptly, and you should scan the paper not just once, but with perseverance. Sometimes the information provided in an advertisement may give you a clue by which you may make the more direct approach described above; you may thus be able to locate an unadvertised position.

PROFESSIONAL ASSOCIATIONS

Scientific societies take an interest in helping employers and prospective employees get together. This interest may be implemented in one or more of several different ways: by advertisements in the official journal of the society, by maintaining placement services in the society's office, or by arranging job interviews at the annual meeting of the organization.

Before going into this in more detail, perhaps it would be helpful to say something about scientific societies in general, and biological associations in particular. There is a great variety of them, and most professional biologists will find it to their advantage to belong to a society that expresses their particular interest; indeed, it is quite the usual thing for a bioscientist to belong to more than one. The societies serve important functions both social and in communications. Most of them publish official journals that help their readers keep up with developments in their field but also keep contact with each other. These functions also are aided by the meetings held on an annual basis, and sometimes more frequently.

Each specialty in bioscience boasts its own society—such organizations as American Association of Anatomists, American Society for Microbiology, and Botanical Society of America. The societies also band together both on a national and an international scale. The qualifications for membership in the various societies vary widely; in some cases, it is only necessary for one to express an interest in the society and pay the dues; others admit to membership only those who have demonstrated ability to conduct and publish results of research in a particular field. Some have more than one category of membership, welcoming not only established professionals but also beginners, with special provision for students.

Science transcends international boundaries, so many of the national societies belong to International Unions, and these in turn make up one grand overall organization called the International Council of Scientific Unions (ICSU). The council's function is to promote meetings in which scientists of various nations get together and exchange information on their latest research findings. On a national scale, the American Association for the Advancement of Science (AAAS) includes scientists of all disciplines, including bioscience. In many states or other local areas there is also a local Academy of Science. The AAAS is an association of many scientific societies, both physical and life sciences, but the individual scientist is also a member of AAAS, and indeed nonscientists or those not members of any constituent society are welcomed as members of AAAS. Annual meetings of AAAS give beginners an opportunity to meet established scientists and to obtain job interviews.

Associations of biological societies of special interest to biologists are the American Institute of Biological Sciences (AIBS) and the Federation of American Societies for Experimental Biology (FASEB). These two large associations are supplemented by societies that belong to no federation. FASEB is primarily of interest to those engaged in research in a biomedical science, as will become apparent by the following constituent societies: American Physiological Society, American Society for Biochemistry and Molecular Biology, American Society for Pharmacology and Experimental Therapeutics, American Association of Pathologists, American Institute of Nutrition, American Association of Immunologists, and American Society for Cell Biology. FASEB does not accept individuals as members, but one is a member of FASEB only by virtue of membership in one or more of the constituent societies. However, nonmembers, especially students, are welcome at meetings, they may present papers, and—most importantly for our present discussion—FASEB operates a placement service that maintains a list of employers with vacancies and those looking for jobs. The list is published annually. In addition, interviews are scheduled between employers and prospective employees.

The largest organization of biologist is AIBS. It is a federation of biological societies, but unlike FASEB, it also accepts individual members. AIBS merits the support of all biologists, and indeed its goal is to enroll all professional biologists. AIBS has provided consulting services for many governmental agencies and private foundations; it has served such groups as the Environmental Protection Agency, the National Science

Foundation, and the Food and Drug Administration. AIBS keeps up-to-date records of biographical and professional information concerning bioscientists and publishes periodicals that keep members and subscribers informed on current interests of biology and biological education.

The Education Committee of AIBS is of special importance to students and teachers or to anyone contemplating a biological career. This committee is actively engaged in career guidance and handles many thousands of requests for career information. These requests come from students, faculty, guidance counselors, and others. The office has sponsored and guided bioscience curriculum studies that have largely revolutionized the teaching of biology in high schools and colleges. Its activities include the preparation and updating of a free brochure, *Careers in Biology,* as well as aiding in the dissemination of career booklets prepared by the individual societies that make up the AIBS. AIBS also charters student chapters in high schools and colleges throughout the country. Participation in a student chapter provides an opportunity for students to maintain contact with biologists and biology professionals all over the country and to gain insight into what biology as a profession is all about.

AIBS maintains a placement service that supplies lists to candidates and to employers so that they can meet by correspondence or for personal interviews at the annual convention. Each year, many biologists, especially beginners, get to meet prospective employers this way. Either FASEB or AIBS will be glad to answer any questions regarding careers or employment at these addresses: FASEB Placement Service, 9650 Rockville Pike, Bethesda, Maryland 20814; AIBS Education Committee, 730 Eleventh Street NW, Washington, DC 20001.

NETWORKING

As you can see, there are many different ways for a biologist to look for a job. Perhaps the most important of all is personal contact. This is especially true as one advances in the profession. During the early stages of a biologist's career, as the student is about to graduate from college, the professors with whom one has been studying will have received notices of fellowships, graduate assistantships, and various teaching, service, or research jobs available to graduates. In many departments, such notices are regularly posted on the bulletin boards;

there also will be some on the professor's desks. Some of these vacancies will be available to those just completing study for a baccalaureate degree; others will be for those about to receive the doctorate. Either way, there is a good chance that one's professor is personally acquainted with the prospective employer. In that case, it is more than likely that the personal recommendation of the professor will carry more weight with the employer than any other single document or record the candidate can produce.

The department head who has a vacancy on the staff for a young person will write to the schools that operate training programs, but will inquire of friends, "Do you know a good man or woman who could fill this position?" At the larger scientific meetings, the corridors are filled with people on the alert for job opportunities, and often with others seeking applicants. If you are one of those looking for a job—or if you already have one but think there might be something better—you will find that as your experience and reputation grow, you will be more and more likely to be one of those sought by employers. Biologists are not a restless lot, but there is some ebb and flow of people among jobs.

After you become known to other biologists as someone whose work merits attention, you may be given chances at more jobs than you can ever fill. When that happy day arrives, the attention paid to you by prospective employers not only gives a lift to your ego, but in many cases opens the eyes of your present boss to your true worth so that you are rewarded with a raise and a promotion to keep you from moving. Then you might join that large group of scientists who like their location so much that they spend practically their entire working lives on one campus or in one laboratory.

PUBLICATIONS AND WEBSITES

Biotechnology: the Choice is Your Future: A Resource Guide. Washington, DC: Biotechnology Industry Organization. Download from www.gene.com/ae/AB/CC/index.html
A comprehensive list of entry-level jobs in biotech industry.

Brown, Sheldon S. *Opportunities in Biotechnology Careers.* Lincolnwood, IL: VGM Career Books, 1990.

Caldwell, Carol Coles. *Opportunities in Nutrition Careers.* Lincolnwood, IL: VGM Career Books, 1992.

Camenson, Blythe. *Opportunities in Museum Careers.* Lincolnwood, IL: VGM Career Books, 1996.

Camenson, Blythe. *Opportunities in Zoo Careers.* Lincolnwood, IL: VGM Career Books, 1997.

"Chapter Two: Changes in the Classroom" and "Chapter Three: The Laboratory Experience" in *Beyond Biology 101: The Transformation of Biology Education.* Chevy Chase, MD: Howard Hughes Medical Institute, 1996. www.hhmi.org/BeyondBio101

Chirico, JoAnn. *Opportunities in Science Technician Careers.* Lincolnwood, IL: VGM Career Books, 1996.

Dehner, Mary. *How to Move from College into a Secure Job.* Lincolnwood, IL: VGM Career Books, 1993.

Dolber, Roslyn. *College and Career Success for Students with Learning Disabilities.* Lincolnwood, IL: VGM Career Books, 1996.

Easton, Thomas. *Careers in Science.* Lincolnwood, IL: VGM Career Books, 1996.

Eberts, Marjorie and Margaret Gisler. *How to Prepare for College.* Lincolnwood, IL: VGM Career Books, 1990.

Education and Training Program in Oceanography and Related Fields. Washington, DC: Marine Technology Society. Phone: (202) 775-5966.

Fanning, Odom. *Opportunities in Environmental Careers.* Lincolnwood, IL: VGM Books, 1995.

Gable, Fred B. *Opportunities in Pharmacy Careers.* Lincolnwood, IL: VGM Career Books, 1994.

Garner, Harry. *Careers in Horticulture and Botany.* Lincolnwood, IL: VGM Career Books, 1997.

"Genetics: the Field of the Future." *Science,* 257:1735. Sept. 18, 1992, or download from www.faseb.org/genetics/gsa/career

Goldberg, Jan. *Opportunities in Horticulture Careers.* Lincolnwood, IL: VGM Career Books, 1995.

Goldberg, Jan. *Opportunities in Research and Development Careers.* Lincolnwood, IL: VGM Career Books, 1997.

Gould, Jay and Wayne Losan. *Opportunities in Technical Writing and Communications Careers.* Lincolnwood, IL: VGM Career Books, 1994.

Green, Marianne. *Internship Success.* Lincolnwood, IL: VGM Career Books, 1997.

Guidelines for Ethical Conduct in the Care and Use of Animals. Hyattsville, MD: American Psychological Association. Write P.O. Box 2710 or download from Website www.apa.org/science/anguide.html

Hoyt, Douglas B. *How to Start and Run a Successful Independent Consulting Business.* Lincolnwood, IL: NTC Career Books, 1996.

Jackson, Acy. *How to Prepare Your Curriculum Vitae.* Lincolnwood, IL: VGM Career Books, 1996.

Kacen, Alex. *Opportunities in Paramedical Careers.* Lincolnwood, IL: VGM Career Books, 1994.

Karni, Karen. *Opportunities in Medical Technology Careers.* Lincolnwood, IL: VGM Career Books, 1996.

Kelsey, Jane. *VGM's Career Portraits in Science.* Lincolnwood, IL: VGM Career Books, 1996.

Kennedy, Joyce Lain. *Career Book.* Lincolnwood, IL: VGM Career Books, 1997.

Kisanne, Sharon F. *Career Success for People with Physical Disabilities.* Lincolnwood, IL: VGM Career Books, 1997.

Kramer, Marc. *Power Networking.* Lincolnwood, IL: VGM Career Books, 1997.

Lussier, Donald and Tom Noteman. *Job Search Secrets.* Lincolnwood, IL: VGM Career Books, 1997.

Marier, Patty and Jan Bailey Mattia. *Choosing a Career Made Easy.* Lincolnwood, IL: VGM Career Books, 1997.

Marier, Patty and Jan Bailey Mattia. *Job Interviews Made Easy.* Lincolnwood, IL: VGM Career Books, 1997.

Marier, Patty and Jan Bailey Mattia. *Networking Made Easy.* Lincolnwood, IL: VGM Career Books, 1997.

Matthews, John R. *The Beginning Entrepreneur.* Lincolnwood, IL: VGM Career Books, 1993.

"Not So Wild a Dream" and "The Xavier Experience." Chevy Chase, MD: The Howard Hughes Medical Institute. (301) 215-8500 (Two free videos that explore ways of attracting minorities to science and helping them succeed.)

On Being a Scientist: Responsible Conduct in Research. Washington, DC: National Academy Press, 1994.

Paradis, Adrian A. *Opportunities in Part-Time and Summer Jobs.* Lincolnwood, IL: VGM Career Books, 1997.

Resource Directory of Scientists and Engineers with Disabilities, Third Edition. Washington, DC: American Association for the Advancement of Science.

"Resources for Teaching Integrity in Scientific Research." Comprehensive bibliography and videotape case studies are available from American Association for the Advancement of Science Directorate for Science and Policy Programs. Website: www.aaas.org.spp.video.resource

Resumes for Science Careers. Lincolnwood, IL: VGM Career Books, 1997.

Riley, Margaret, Frances Roehm, and Steve Oserman. *The Guide to Internet Job Searching.* Lincolnwood, IL: VGM Career Books, 1997.

Rogers, Carla. *How to Get into the Right Medical School.* Lincolnwood, IL: VGM Career Books, 1996.

Rowh, Mark. *Slam Dunk Cover Letters.* Lincolnwood, IL: VGM Career Books, 1997.

Rubinfeld, William A. *Planning Your College Education.* Lincolnwood, IL: VGM Career Books, 1997.

Sacks, Terence J. *Opportunities in Physician Assistant Careers.* Lincolnwood, IL: VGM Career Books, 1995.

Shingleton, John D. *Job Interviewing for College Students.* Lincolnwood, IL: VGM Career Books, 1995.

Snook, I. Donald. *Opportunities in Health and Medical Careers.* Lincolnwood, IL: VGM Career Books, 1997.

White, William C. and Donald Collins. *Opportunities in Farming and Agriculture.* Lincolnwood, IL: VGM Career Books, 1995.

Willie, Christopher. *Opportunities in Forestry Careers.* Lincolnwood, IL: VGM Career Books, 1998.

FAVORITE WEBSITES

Careers for Biology Majors: www.as.udayton.edu/bio/career
Two-hundred links to sources of information about careers in biology; also has a section on FAQ's of Life.

Hot Biology Web Sites: www.zoo.utoronto.edu/zooweb/able/notsites/hotsites/htm
Links to many more sources of information about biology including organizations, virtual courses, a dictionary, and so forth.

The Virtual Career and Education Center: www.petersons.com
You can create lists of educational institutions offering the major of your choice and make a list of potential employers from the database of the publisher of the Peterson guides.

ORGANIZATIONS

American Association for the Advancement of Science
1200 New York Avenue NW
Washington DC 20005
Phone: (202) 326-6400
Website: www.aaas.org

> A wealth of information on science careers and education and publishes *Science.*

American Association of Botanical Gardens and Arborita
786 Church Road
Wayne, PA 19087
Website: www.mobot.org/aanga

> Website lists resource center materials on careers, educational programs, volunteer and internship opportunities in botanical and zoologic gardens.

American Fisheries Society
5410 Grosvenor Lane, Suite 110
Bethesda, MD 20814
Phone: (301) 897-8616
Website: www.ESD.ORNL.GOV/societies/AFS

> Offers a certification program for fisheries professionals.

American Indian Science and Engineering Society
5661 Airport Boulevard
Boulder, CO 80301-2339
Phone: (303) 939-0023
Website: http://www.colorado.edu/AISES

> Provides scholarships and information for Native American students in the United States and Canada.

American Institute of Biological Sciences (AIBS)
  1313 Dolly Madison Boulevard, Suite 402
  McLean, VA 22101
  Phone: (800) 992-2427
  Website: www.aibs.org

  Maintains lists of professional opportunities for job seekers.

American Pharmaceutical Association
  2215 Constitution Avenue NW
  Washington, DC 20037
  Phone: (202) 628-4410

American Physiological Association
  9560 Rockville Pike
  Bethesda, MD 20814
  Phone: (310) 530-5500

American Society of Biochemistry and Molecular Biology
  9650 Rockville Pike
  Bethesda, MD 20814
  Phone: (301) 530-7153

American Society of Human Genetics
  9650 Rockville Pike
  Bethesda, MD 20814
  Phone: (301) 571-1825

  Publishes guides to careers and postgraduate programs.

American Society of Limnology and Oceanography
  Virginia Institute of Marine Science
  College of William and Mary
  Route 1208
  Gloucester Point, VA 23062

  Detailed information on jobs, employment outlook, working conditions,
    earnings, colleges, and universities.

American Society for Microbiology
  1325 Massachusetts Avenue NW
  Washington, DC 20005-4171
  Website: www.asmusa.org

  Website contains excerpts from *Your Career in Microbiology: Unlocking the
    Secrets of Life,* which also can be ordered from ASM.

American Zoo and Aquarium Association
    7970-D Old Georgetown Road
    Bethesda, MD 20814-2493
    Website: www.aza/career.htm

    Information on types of jobs, salaries, and so forth.

Association of American Medical Colleges
    2450 N Street NW
    Washington, DC 20037-1126
    Phone: (202) 828-0584
    Website: www.aamc.org.

    Working to increase minority students enrolled in medical schools.

Association for Biology Laboratory Education
    Department of Biology
    Yale University
    P.O. Box 208104
    New Haven, CT 06520-8104
    Phone: (203) 432-3864
    Website: www.zoo.utoronto.ca/~able

    College and university professors share successful labs ideas.

Association of Systemics Collections
    1725 K Street NW
    Washington, DC 20006-1401
    Phone: (202) 835-9050
    Website: www.ascoll.org

    Information about career opportunities.

Association of Women in Science
    1200 New York Avenue, NW
    Washington, DC 20005
    Phone: (202) 326-8940
    Website: www.awis.org.

    For women in all areas of science and technology.

ASPIRA, Inc
    1444 I Street NW, 8th floor
    Washington, DC 20005
    Phone: (202) 326-8940
    E-mail: awis@awis.org.

    Educational opportunities for Puerto Rican and Latino youth.

Bioethics Centre
University of Alberta, Room 222
Aberhart Nurses Residence
8220-114 Street
Edmonton, Alberta, T6G 2J3
Canada
Phone: (403) 492-6676
Website: www.gpu.srv.UAlberta.Ca

Promotes awareness of bioethical issues in biomedicine and publishes the
Bioethics Bulletin.

Botanical Society of America
Ohio State University
1735 Neil Avenue
Columbus, OH 43210
Phone: (614) 292-3519
Website: www.ou.edu/cas/botany-micro/careers

Write for or download from web "Careers in Botany: a Guide to Working
with Plants."

Colorado State University
Department of Entomology
Website: www.colostate.edu

Maintains a current list of job opportunities in entomology.

Ecological Society of America
2010 Massachusetts Avenue NW, Suite 400
Washington, DC 20036
Phone: (202) 833-8773

Website: www.sdsc.edu/1/SDSC/Research/CompBio/ESA/ESA.html

Entomological Society of America
9301 Annapolis Road
Lanham, MD 20706
Website: www.entsoc.org.

List of institutions with programs, certification program, job opportunities,
and chat.

Federation of American Societies for Experimental Biology (FASEB)
9650 Rockville Pike
Bethesda, MD 20814-3998
Phone: (301) 530-7020
Website: www.faseb.prg/careers

Placement and career info for job seekers.

Mycological Society of America
Websites: www.erin.utoronto.ca/soc/msa or muse.bio.cornell.edu/~fungi/findex

Everything you always wanted to know about mycology!

National Agricultural Chemicals Association
1155 Fifteenth Street, NW
Washington, DC 20005

Career information for entomologists.

National Association of Biology Teachers
11230 Roger Bacon Drive, #19
Reston, VA 22090
Phone: (703) 471-1134
Website: www.nabt.org

Publishes *The American Biology Teacher.*

National Pest Control Association
8100 Oak Street
Dunn Loring, VA 22027

Career information for entomologists and pest control technicians.

National Science Foundation
4201 Wilson Boulevard
Arlington, VA 22230

Facilitation awards for handicapped scientists: (703) 306-1636.

National Science Teachers Association
1840 Wilson Boulevard
Arlington, VA 22201
Phone: (703) 312-9232
Website: www.nsta.org

Publishes *Journal of College Science Teaching.*

Scripps Institution of Oceanography
University of California, San Diego
Website: www-siograddept.ucsd,edu/web/To_Be_A_Marine_Biologist

A graduate student frequently asked questions about preparing for a career in oceanography.

Sea World and Busch Gardens at Website: www.seaworld.org/Careers/careerinfo.html.

Contains information about careers in zoos or aquariums and about camp for high school students to try out these careers.

U.S. Environmental Protection Agency (EPA)
Office of Human Resources
Website: www.epa.gov/epahrist

Lists employment and internship opportunities.

U.S. Fish and Wildlife Service
Dept. of the Interior
1849 C Street NW
Washington, DC
Phone: (703) 358-2120
Website: www.fws.gov

Comprehensive list of careers for biologists.

Whitaker Foundation
1700 North Moore Street, Suite 2200
Rosslyn, VA 22209
Phone: (703) 528-2430
Website: www.whitaker.org.

Private, nonprofit foundation that supports research and education in biomedical engineering; career information is available on website or write for publications.

GLOSSARY

AAAS.   The American Association for the Advancement of Science.

Agricultural science.   Any of several applied biological sciences seeking to discover knowledge about plants and animals and apply the discoveries to the needs of humankind for food or fiber.

Agronomy.   Scientific agriculture; the application of plant sciences and soil science to the raising of crops and management of soil.

AIBS.   The American Institute of Biological Sciences.

Algology.   A branch of systematic botany dealing with the algae, simple green plants comprising the seaweeds and similar forms of fresh water and damp places.

Anatomy.   The biological science dealing with the structure of a plant or an animal; used more or less interchangeably with the word *morphology.*

Animal science.   A term often used more or less synonymously with *animal husbandry;* the branch of applied biology relating to the breeding, nutrition, physiology, and diseases of domestic animals, especially those of importance to agriculture.

Aquaculture.   The practical application of biological knowledge to the cultivation of the growth of organisms, usually fish, oysters, or shrimp, in water; analagous to agriculture on land; if sea water is involved, it is often called mariculture.

Aquatic biology.   Biological science, either basic or applied, concerned with biological events or organisms in water; if sea water is involved, it is usually called marine biology and may be considered to be a branch of oceanography; as sometimes used, the term *aquatic biology* refers to fresh water and is a branch of limnology.

Bacteriology. The scientific study of the bacteria; a branch of microbiology.

Biochemistry. The scientific study of the chemistry of living things; the study of living matter or of substances derived from living matter whether plant, animal, or microorganism, by the use of the methods of chemistry. Synonyms: biological chemistry, physiological chemistry.

Biology. A broad and all-inclusive term for any scientific study of life; all the sciences discussed in this book are branches of biology.

Bionics. The application of biological principles to the study and design of engineering systems.

Bionomics. See *ecology.*

Biophysics. A branch of biology that uses the tools and concepts of physics to study living matter.

Biosystematics. The branch of biology dealing with the description, classification, and naming of plants and animals (see *taxonomy*), together with whatever features distinguish one species from another, whether anatomical, biochemical, or ecological, and any information enabling the bioscientist to place an organism in its place in the evolutionary process; also called systematics or systematic biology.

Biotechnology. The application of biological principles to practical objectives, especially as illustrated by such procedures as genetic engineering, cell fusion, and embryo transfer.

Botany. The scientific study of plants; a general term including many biological disciplines, as applied to plants.

Bryology. A branch of systematic botany concerned with mosses and liverworts.

Cytology. The scientific study of cells, including structure, formation, biochemistry, and functions.

Dendrology. The branch of botany concerned with trees and shrubs.

Ecology. The scientific study of the relationship of an organism—plant, animal, or human—to its environment; sometimes called bionomics.

Embryology. The study of organisms in their earliest stages of development; in zoology, before hatching (oviparous animals) or in rudimentary beginning stages (viviparous animals); in botany, the rudimentary plant contained in the seed.

Endocrinology. A branch of physiology concerned with the endocrine glands or other structures that release substances exerting physiological effects at sites other than their points of origin.

Entomology. The scientific study of insects.

Ethology.   The study of animal behavior by observing normal activities; for example, the imitation by a member of a species of the actions of others (original meaning: the portrayal of character by mimicry).

Evolution.   The changes in inherited characteristics of a species so that descendents may differ morphologically or physiologically from their ancestors.

FASEB.   The Federation of American Societies for Experimental Biology.

Forestry.   The branch of applied science concerned with all aspects of the planting and care of forest trees.

Genetic engineering.   The alteration of genetic material by chemical or physical means to produce an intentional change in inherited characteristics.

Genetics.   The science of heredity; concerned with all aspects of the biological processes involved in inheritance or the passing on of characteristics from one generation to the next.

Gnotobiotics.   A branch of biology dealing with breeding and growing animals in a germ-free environment.

Hematology.   The study of blood, including both cellular and fluid parts.

Herpetology.   A branch of systematic zoology dealing with reptiles and amphibians.

Histology.   A study of the microscopic structure of the tissues of plant or animal.

Horticulture.   A study of the cultivation of garden plants.

Hybridoma.   An artificial organism or new type of cell formed by fusion of two different types of cells.

Ichthyology.   A branch of systematic zoology dealing with fishes.

ICSU.   The International Council of Scientific Unions.

Immunology.   The scientific study of the bodily processes involved in immunity to disease and similar biochemical reactions.

Limnology.   The scientific study of bodies of fresh water; includes physical, geographic, as well as biological aspects; the corresponding study of the ocean is oceanography.

Mammology.   The branch of zoology dealing with mammals, animals that suckle their young.

Mariculture.   See *aquaculture.*

Marine biology.   See *aquatic biology.*

Microbiology.   The scientific study of microscopic organisms.

Molecular biology.   The study of the molecules that direct molecular processes in cells; a specialty within biochemistry.

Morphology. The study of the structure of an organism without regard to function; same as anatomy.

Mycology. A branch of systematic botany dealing with fungi—that is, yeasts, molds, and mushrooms.

NABT. The National Association of Biology Teachers.

Oceanography. The scientific study of the oceans.

Organism. Any living form, whether an animal, a plant, a bacterium, or other.

Ornithology. A branch of systematic zoology dealing with birds.

Parasitology. The scientific study of parasites—that is, of plants or animals that obtain their nutriment by living in or upon other living bodies.

Pathology. The scientific study of disease, including its origin, nature, and effects upon the body.

Pharmacology. The scientific study of the effects of foreign chemical substances upon bodily functions; the study of drugs and their effects on the body.

Physiology. The scientific study of functions of living organisms or of any of their parts.

Phytopathology. The study of the diseases of plants.

Plankton. Small plants (phytoplankton) or animals (zooplankton) that float or drift about in the water, especially those near the surface; plankton are at the bottom of the food pyramid—they are eaten by larger creatures.

Plant pathology. See *phytopathology.*

Plant physiology. The study of functions of plant organisms or of any of their parts.

Protozoology. The scientific study of one-celled animals.

Systematics; Systematic Biology. See *biosystematics.*

Taxonomy. Description, classification, and naming of organisms; an integral part of biosystematics.

Technician. In the biology laboratory, a technician is a laboratory worker who is skilled in the art and technique of biology and is a valuable assistant to the professional scientist in charge of the laboratory, and who may supervise other technicians, but who lacks the theoretical background or experience to have full charge of complex operations such as research projects.

Technologist. A technician; as commonly used, a technologist may have a higher degree of formal training than a technician (as these terms are officially employed by the American Society of Clinical Pathologists, for example).

Toxicology.   The scientific study of poisons, their detection, effects, and antidotes—a branch of pharmacology.

Virology.   The study of viruses and of the diseases caused by them.

Virus.   An infectious agent requiring living cells for propagation; viruses are smaller than bacteria.

Zoology.   The scientific study of animals; a general term including many biological disciplines as applied to animals.

Zoonosis.   A disease of animals transmissable to humans; plural, *zoonoses.*